Mathematics

with Business Applications

Fourth Edition

Problems and Simulations

Walter H. Lange
Temeoleon G. Rousos

New York, New York   Columbus, Ohio   Woodland Hills, California   Peoria, Illinois

A Division of The McGraw-Hill Companies

CONTENTS

Glencoe/McGraw-Hill
A Division of The McGraw-Hill Companies

Copyright © 1998, 1993 by Glencoe/McGraw-Hill. All rights reserved. Copyright 1986, 1984, 1981 by Houghton-Mifflin. Permission is granted to reproduce the material contained herein on the condition that such material be reproduced only for classroom use; be provided to students, teachers, and families without charge; and be used solely in conjunction with *Mathematics with Business Applications*, fourth edition. Any other reproduction, for use or sale, is prohibited without prior written permission from the publisher.

Printed in the United States of America.

Send all inquiries to:
Glencoe/McGraw-Hill
21600 Oxnard Street, Suite 500
Woodland Hills, CA 91367

ISBN 0-02-814732-4

4 5 6 7 8 9 10 066 04 03 02 01 00

Name ______________________ Date ______________

REVIEW OF FUNDAMENTALS

WORKSHOP 1: Writing and Rounding Numbers

The place-value chart gives the value of each digit in the number 4532.869. The place-value chart can help you write numbers.

32.86 is thirty-two and eighty-six hundredths.

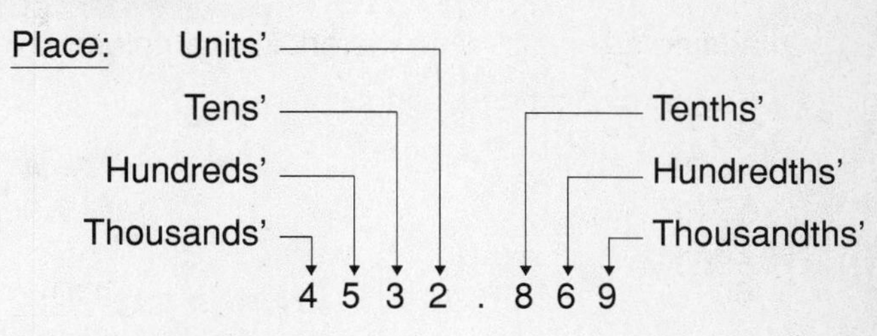

Place value is also used in rounding. Round 7437 to the nearest hundred.

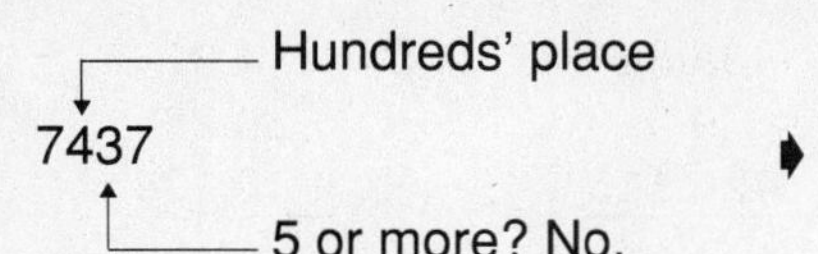

Do not change.

7437

Drop the final digits and replace with zeros.

The answer is 7400.

Write the number.

1. Four thousand two hundred twenty-six ______

2. Five hundred three and fifteen hundredths ______

Write in word form.

3. 723 ______

4. 3634.68 ______

______

Round to the place value shown.

Nearest hundred. **5.** 5372 ______ **6.** 4331 ______

Nearest tenth. **7.** 0.3815 ______ **8.** 10.091 ______

WORKSHOP 2: Comparing Numbers

Which number is greater, 586.900 or 586.679?

To compare the numbers, compare each digit beginning at the left.

5	8	6	.	9	0	0
5	8	6	.	6	7	9
same	same	same		9 is greater than 6. So 586.900 is greater than 586.679.		

Which number is greater?

9. 221 or 214 ______ **10.** 4379 or 4397 ______

11. 15.7 or 15.712 ______ **12.** 12.124 or 12.13 ______

Copyright © by Glencoe Division

Name ____________________ Date ____________

REVIEW OF FUNDAMENTALS

WORKSHOP 3: Adding Decimals

When adding decimals, write the addition problem in vertical form, being sure to line up the decimal points.

14.93 + 17.29 + 28.64

```
  14.93
  17.29
+ 28.64
  60.86
```

36.4 + 2.96 + 7 + 110.513

```
   36.4            36.400
    2.96     ➧      2.960
    7.              7.000
+ 110.513      + 110.513
                 156.873
```

1.
```
  26.24
  36.37
+ 39.88
```

2.
```
  8.516
  9.220
+ 3.897
```

3.
```
    9.2
   42.76
+ 126.923
```

4.
```
   38.34
    9.0
+ 426.115
```

5.
```
   4.37
  29.26
+  0.124
```

6. 86.42 + 29.45 + 15.01 + 43.82 ____________

7. 0.95 + 8.91 + 26.421 + 18.2 ____________

8. 8.2 + 19.81 + 435.67 + 2.1 ____________

9. 45.23 + 0.21 + 15 + 0.921 ____________

10. $4.19 + $47.21 + $.16 + $4.24 ____________

WORKSHOP 4: Subtracting Decimals

When subtracting decimals, write the subtraction problem in vertical form, being sure to line up the decimal points.

87.36 − 24.14

```
  87.36
− 24.14
  63.22
```

129.7 − 18.56

```
  129.7    ➧    129.70
−  18.56      −  18.56
                111.14
```

11.
```
  4.93
− 2.81
```

12.
```
  181.52
−  11.41
```

13.
```
  847.86
− 121.92
```

14.
```
  234.07
− 119.94
```

15.
```
  242.19
−  98.38
```

16. 527.8 − 114.6 ____________

17. 44.39 − 1.85 ____________

18. $572.11 − $328.64 ____________

19. $1276.98 − $1053.73 ____________

20. $100.00 − $4.72 ____________

21. $378.22 − $144.84 ____________

Copyright © by Glencoe Division

Name ____________________ Date ____________

REVIEW OF FUNDAMENTALS

WORKSHOP 5: Multiplying Decimals

When multiplying decimals, multiply as if the decimal numbers were whole numbers. Then count the total number of decimal places in the factors. This number will be the number of decimal places in the product.

```
    16.4  ← factor            16.4  ←    1 decimal place
 ×  0.28  ← factor         ×  0.28  ← + 2 decimal places
    1312                      1312
    328                       328
    4592  ← product           4.592 ←    3 decimal places
```

1. $\begin{array}{r} 23.4 \\ \times\ 2.6 \\ \hline \end{array}$ **2.** $\begin{array}{r} 25.5 \\ \times\ 6.4 \\ \hline \end{array}$ **3.** $\begin{array}{r} 39.8 \\ \times\ 0.27 \\ \hline \end{array}$ **4.** $\begin{array}{r} 76.4 \\ \times\ 0.14 \\ \hline \end{array}$ **5.** $\begin{array}{r} \$14.28 \\ \times\ 132 \\ \hline \end{array}$

6. How much will you pay for 14.8 gallons of diesel fuel at $1.39 per gallon? Round to the nearest cent. ____________

WORKSHOP 6: Dividing Decimals

```
            231 ← quotient
divisor → 3 ) 693 ← dividend
```

When dividing decimals, if there is a decimal point in the divisor, you must move it to the right to make the divisor a whole number. Then move the decimal point in the dividend to the right the same number of places you moved the decimal point in the divisor. Then divide as with whole numbers.

```
                                   6.37
7.23 ) 46.0551              723 ) 4605.51
                                  4338
                                   2675
                                   2169
                                    5061
                                    5061
```

Divide. Round to the nearest hundredth or to the nearest cent.

7. 36) 262.8 **8.** 6.2) 22.444 **9.** 0.15) .8879 **10.** 44) $968.35

11. Herman Holdeman drove 591 miles on 15.4 gallons of gasoline. How many miles per gallon did he get? ____________

Copyright © by Glencoe Division

Name ______________________________ Date ______________

REVIEW OF FUNDAMENTALS

WORKSHOP 7: Writing Fractions as Decimals and Decimals as Fractions

Any fraction can be renamed as a decimal and any decimal as a fraction. To change a fraction to a decimal, divide the numerator by the denominator.

$\frac{5}{8} \rightarrow$

```
     0.625
8 ) 5.000
    48
     20
     16
      40
      40
```

$\frac{17}{50} \rightarrow$

```
      0.34
50 ) 17.00
     150
      200
      200
```

To rename a decimal as a fraction, name the place value of the digit at the far right. This is the denominator of the fraction.

$0.87 = \frac{87}{100}$ $\quad$ $0.08 = \frac{8}{100} = \frac{2}{25}$ $\quad$ $6.375 = 6\,\frac{375}{1000} = 6\frac{3}{8}$

Write as decimals. Round answers to the nearest hundredth.

1. $\frac{2}{5}$ ________ **2.** $\frac{1}{4}$ ________ **3.** $\frac{3}{8}$ ________ **4.** $\frac{7}{10}$ ________

5. $4\frac{5}{8}$ ________ **6.** $5\frac{1}{5}$ ________ **7.** $8\frac{1}{9}$ ________ **8.** $2\frac{3}{25}$ ________

Write as fractions in lowest terms.

9. 0.4 ________ **10.** 0.07 ________ **11.** 0.83 ________ **12.** 3.23 ________

13. 1.25 ________ **14.** 2.45 ________ **15.** 7.830 ________ **16.** 9.75 ________

WORKSHOP 8: Writing Decimals as Percents and Percents as Decimals

Percent means divided by 100. To write a percent as a decimal, move the decimal two places to the left and drop the % sign.

15% = 15. = 0.15 $\quad$ 56.2% = 56.2 = 0.562 $\quad$ 134% = 134. = 1.34

To write a decimal as a percent, move the decimal two places to the right and add a % sign.

0.25 = 0.25 = 25% $\quad$ 3.42 = 3.42 = 342% $\quad$ .005 = .005 = 0.5%

Write as decimals.

1. 19% ________ **2.** 25% ________ **3.** 47% ________ **4.** 8% ________

5. 244% ________ **6.** 104% ________ **7.** 39.8% ________ **8.** 17.2% ________

Write as percents.

9. 0.82 ________ **10.** 0.51 ________ **11.** 0.019 ________ **12.** 3.17 ________

Copyright © by Glencoe Division

Name ______________________ Date ______________

REVIEW OF FUNDAMENTALS

WORKSHOP 9: Finding a Percentage

In mathematics, of means "times" and is means "equals."

35% of 80 is what number?

$35\% \times 80 = n$ ← Let "n" stand for the unknown number.
$0.35 \times 80 = n$ ← Change the percent to a decimal.
$28 = n$ Multiply.
$35\% \text{ of } 80 = 28$ ← Write the answer.

Find the percentages.

1. 20% of 60 ________ **2.** 60% of 50 ________ **3.** 35% of 86 ________

4. 9.5% of 44 ________ **5.** 18.2% of 260 ________ **6.** 17.3% of 85 ________

7. 6% of 856 ________ **8.** 0.6% of 15.6 ________ **9.** 125% of 60 ________

10. 18% of $70 ________ **11.** $8\frac{1}{2}$% of $400 ________ **12.** 5.5% of $268 ________

13. 65% of the 880 students at Morton High School take mathematics. How many students take mathematics? ________

14. Sales tax: 6.5% of your $8.98 purchase. What is the sales tax? ________

WORKSHOP 10: Average (Mean)

The average, or mean, of two or more numbers is the sum of the numbers divided by the number of items added.

To find the mean of $42.90, $64.20, $37.52, and $81.86:

$$\frac{\$42.90 + \$64.20 + \$37.52 + \$81.86}{4} = \frac{\$226.48}{4} = \$56.62$$

Find the mean for each group.

1. 10, 12, 15, 11 ________ **2.** 420, 460, 410, 470 ________

3. 0.5, 0.29, 0.42, 0.36 ________ **4.** $42.50, $42.80, $37.60, $40.20 ________

Use the rates of taxation chart on page 7 for problems 5–7.

5. What is the average general school tax rate for the five school districts? ________

6. What is the average construction school tax rate for the five districts? ________

7. What is the average total school tax for the five school districts? ________

Copyright © by Glencoe Division

Name ______________________ Date ______________

REVIEW OF FUNDAMENTALS

WORKSHOP 11: Elapsed Time

To find elapsed time, subtract the earlier time from the later time.

If you worked from 3:30 P.M. to 5:45 P.M., how long did you work?

$$\begin{array}{r} 5{:}45 \\ -\ 3{:}30 \\ \hline 2\ \text{h}{:}15\ \text{min} \end{array}$$

You worked 2 hours and 15 minutes.

If you worked from 3:30 P.M. to 6:15 P.M., how long did you work? You cannot subtract 30 from 15. Use the fact that 1 hour equals 60 minutes.

$$\begin{array}{rcrcr} 6{:}15 & = & 5{:}15 + {:}60 & = & 5{:}75 \\ -\ 3{:}30 & = & -\ 3{:}30 & = & -\ 3{:}30 \\ \hline & & & & 2\ \text{h}{:}45\ \text{min} \end{array}$$

You worked 2 hours and 45 minutes.

To find elapsed time when the time period spans 1 o'clock, add 12 hours to the later time before subtracting. To find elapsed time from 9:45 A.M. to 2:50 P.M.:

$$\begin{array}{rcrcr} 2{:}50 & = & 2{:}50 + 12{:}00 & = & 14{:}50 \\ -\ 9{:}45 & = & -\ 9{:}45 & = & -\ 9{:}45 \\ \hline & & & & 5\ \text{h}{:}05\ \text{min} \end{array}$$

Find the elapsed time.

1. From 4:30 P.M. to 6:40 P.M. ______________

2. From 2:15 P.M. to 6:50 P.M. ______________

3. From 7:15 A.M. to 12:55 A.M. ______________

4. From 8:45 A.M. to 4:37 P.M. ______________

5. From 2:12 P.M. to 11:10 P.M. ______________

6. From 3:44 P.M. to 10:25 P.M. ______________

7. From 10:18 A.M. to 6:25 P.M. ______________

8. From 7:06 A.M. to 5:01 P.M. ______________

9. Marianne LaRouse worked from 7:15 A.M. to 11:25 A.M., went to lunch, and then worked from 12:09 P.M. to 3:48 P.M. How long did she work? ______________

10. Steve Galvin worked from 7:55 A.M. to 11:45 A.M., went to lunch, and then worked from 12:15 P.M. to 5:10 P.M. How long did he work? ______________

11. Joseph Corey took an airplane that left Toledo at 8:17 A.M. and arrived in New York City at 11:05 A.M. How long was the trip? ______________

12. Nancy Ness took a bus that left Chicago at 8:42 P.M. and arrived in St. Louis at 2:45 A.M. How long was the trip? ______________

13. Nolan Burroughs took an airplane that left Hartford at 5:40 P.M. and arrived in Cleveland at 7:28 P.M. Then he left Cleveland at 8:05 P.M. and arrived in Detroit at 8:35 P.M. What was his total flying time? ______________

Copyright © by Glencoe Division

Name ______________________________ Date ______________

REVIEW OF FUNDAMENTALS

WORKSHOP 12: Reading Tables and Charts

To read a table, find the column containing one of the conditions of the information you are looking for. Look across the row containing the other condition until it crosses the column you found. Read the answer.

What is the total school tax rate for Swanton Township schools?

Find the column headed "Total."

Find the row for Swanton Township.

Read across the Swanton Township row to the "Total" column.

The total school tax rate for Swanton Township is $96.20.

TAXATION PER $1000 OF ASSESSED VALUE			
School District	Purpose		
	General	Construction	Total
Allen Township	$89.55	$4.35	$93.90
Swanton Township	87.00	9.20	96.20
Clay City	86.85	8.70	95.55
Salem Township	77.50	5.30	82.80
Oak Harbor City	78.60	8.90	87.50

1. What is the general school tax rate for Oak Harbor City? ______________

2. What is the construction tax rate for Clay City? ______________

3. Which school district has the lowest total school tax rate? ______________

4. Which school district has the lowest construction school tax rate? ______________

5. Which school district has a general school tax rate of $87.00? ______________

WORKSHOP 13: Constructing Graphs

A bar graph is a picture that graphically displays and compares numerical facts in the form of vertical or horizontal bars.

1. Construct a vertical bar graph of the given data.

Average Price of Unleaded Gasoline	
1987	$0.95
1988	1.00
1989	1.10
1990	1.35
1991	1.15

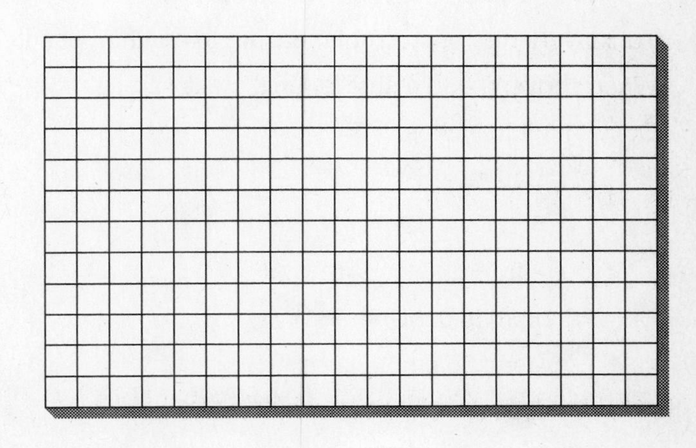

Copyright © by Glencoe Division

Name ______________________ Date ______________

REVIEW OF FUNDAMENTALS

WORKSHOP 13 (continued)

A line graph is a picture used to compare facts over a period of time. It is an excellent way to show trends (increases or decreases).

2. Construct a line graph of the data given on page 7 for unleaded gasoline.

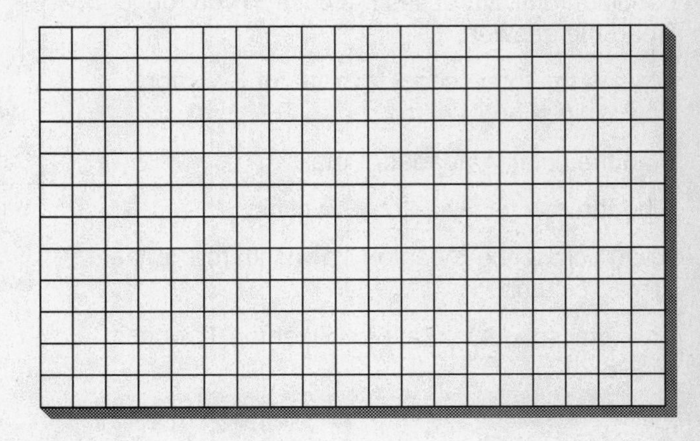

WORKSHOP 14: Units of Measure

Here are the abbreviations and conversions for units of measure in the U.S. Customary System.

Length	Volume	Weight
12 inches (in.) = 1 foot (ft)	2 cups (c) = 1 pint (pt)	16 ounces (oz) = 1 pound (lb)
3 ft = 1 yard (yd)	2 pt = 1 quart (qt)	2000 lb = 1 ton (t)
5280 ft = 1 mile (mi)	4 qt = 1 gallon (gal)	

Here are the symbols and conversions for units of measure in the metric system.

Length	Volume
1000 millimeters (mm) = 1 meter (m)	1000 milliliters (mL) = 1 liter (L)
100 centimeters (cm) = 1 m	**Mass**
1000 m = 1 kilometer (km)	1000 grams (g) = 1 kilogram (kg)

To convert from one unit of measure to another, use the conversion lists above.

When converting to a smaller unit, multiply.

Convert 8 ft to inches.

Use 12 in. = 1 ft

8 ft: 8 × 12 = 96

8 ft = 96 in.

Convert 2.5 m to centimeters.

Use 100 cm = 1 m

2.5 m: 2.5 × 100 = 250

2.5 m = 250 cm

When converting to a larger unit, divide.

Convert 15 qt to gallons.

Use 4 qt = 1 gal

15 qt: 15 ÷ 4 = 3.75

15 qt = 3.75 gal

Convert 4240 mL to liters.

Use 1000 mL = 1 L

4240 mL: 4240 ÷ 1000 = 4.240

4240 mL = 4.240 L

Copyright © by Glencoe Division

Name ______________________ Date ______________

WORKSHOP 14 (continued)

Do the following conversions.

1. 0.5 mi to feet ________

2. 28 qt to gallons ________

3. 9 qt to pints ________

4. 2.2 L to milliliters ________

5. 129 ft to yards ________

6. 1.5 lb to ounces ________

7. 4.5 km to meters ________

8. 6300 lb to tons ________

9. 95 m to kilometers ________

10. 24 c to quarts ________

11. 13 c to pints ________

12. 4240 g to kilograms ________

13. 7.23 kg to grams ________

14. 3242 mm to meters ________

15. Lisa O'Leary jogged around a track 20 times. The distance around the track is 1320 feet. How many miles did she run? ________

WORKSHOP 15: Estimating—Front End

Estimation is a valuable tool. It is used as a quick check of the reasonableness of a calculation. One way to estimate is to add the front-end digits after adjusting the sum of the front-end digits.

Write the estimated answer in the blank and then perform the computations.

1.

```
  3437
  2492
+ 1651   + ______
```

2.

```
  37.39
  16.21
+ 51.63   + ______
```

3.

```
  253
  461
  805
+ 763   + ______
```

4. Estimate the total attendance for five home football games. Game 1: 3214; Game 2: 1374; Game 3: 2472; Game 4: 3010; Game 5: 1861.

Estimate: ________

Actual: ________

5. Estimate the cost of these auto parts. Muffler: $35.25; exhaust pipe: $17.90; tail pipe: $27.35; clamps: $9.90; bracket: $8.32.

Estimate: ________

Actual: ________

Copyright © by Glencoe Division

Name ______________________ Date ______________

REVIEW OF FUNDAMENTALS

WORKSHOP 16: ESTIMATION—ROUNDING

Rounding is used to estimate an answer. Round the numbers to the highest place value, then perform the indicated computation.

Find	Estimate	Find	Estimate
47.7	50	21.2	20
+ 33.2	+ 30	× 5.8	× 6
	80		120

The answer is <u>about</u> 80.

By computation it is 80.9.

The answer is <u>about</u> 120.

By computation it is 122.96.

Estimate first. Then perform the indicated computation.

1. 4780 + 3817

2. 76.7 − 31.8

3. 114.32 × 17.4

4. 476 ÷ 52.4 = ______________

____

5. 24.6% × 397 = ______________

____

6. Fifty-eight people charter a bus for $312. Estimate the amount each person pays, then calculate. ______________

WORKSHOP 17: ESTIMATION—COMPATIBLE NUMBERS

Change the numbers in your problem to numbers that are easy to compute. These are called compatible numbers.

Find	Estimate	Find	Estimate
$\frac{1}{3} \times 11\frac{5}{8}$	$\frac{1}{3} \times 12 = 4$	25.5% of $589	$\frac{1}{4} \times \$600 = \$150$

The estimate is easy since 12 is close to $11\frac{5}{8}$ and is compatible with $\frac{1}{3}$.

The answer is <u>about</u> 4.

By computation it is $3\frac{7}{8}$

The estimate is easy since 25.5% is about $\frac{1}{4}$ and $589 is about 600. 600 and $\frac{1}{4}$ are compatible.

The answer is <u>about</u> $150.

By computation it is $150.20.

Estimate using compatible numbers. Perform the computations.

1. 4124 ÷ 8.1 = ______________

____

2. 21.4% × $496.80 = ______________

____

3. $\frac{5}{16}$ × 911 = ______________

____

4. 68% × $1196 = ______________

____

5. 158.7 ÷ 41.2 = ______________

____

6. Harriet Murdock saves 23% of her paycheck each week. Last week her check was for $279.88. Estimate the amount she saved. ______________

Copyright © by Glencoe Division

Name ______________________ Date ______________

REVIEW OF FUNDAMENTALS

WORKSHOP 18: ESTIMATION—CLUSTERING

When the numbers to be added are close to the same quantity, the sum can be found by clustering.

$6.95	All of the	$19.79	
7.11	numbers cluster	21.49	about $60.00
6.89	around $7.00 so	19.99	
+ 7.19	$7 × 4 = $28.	+ 8.99	+ 9.00
			about $69.00

The answer is about $28.00.

By computation it is $28.14.

The answer is about $69.00.

By computation it is $70.26.

Estimate the sums by clustering. Compute the actual amount.

1.	2.	3.	4.	5.
$14.79	$79.79	$17.97	$49.79	$199.79
14.89	81.49	18.19	52.19	209.19
15.19	78.99	17.79	47.79	189.49
+ 15.29	+ 81.19	+ 4.89	9.79	79.99
			+ 9.89	+ 77.79

Est.

Act.

6. Quick Trip Delivery Service drove the miles indicated: Mon., 489; Tues., 511; Wed., 479; Thur., 494; Fri., 507; and Sat., 487. Estimate and calculate the total miles for the week. ____________

WORKSHOP 19: PROBLEM SOLVING—Four-Step Method

The Four-Step Method

STEP 1: UNDERSTAND — What is the problem? What is given?

STEP 2: PLAN — What do you need to do to solve the problem?

STEP 3: WORK — Carry out the plan. Do any necessary calculations.

STEP 4: ANSWER — Is your answer reasonable? Did you answer the question?

EXAMPLE A commercial building is being remodeled. It will take 4 rough carpenters 7 days to frame out the building. Each rough carpenter works 8 hours a day at $18 per hour. How much will it cost the remodeler for the rough carpenters?

SOLUTION

STEP 1. UNDERSTAND — 4 rough carpenters, 7 days, 8 hours per day, $18 per hour
The cost per day for one rough carpenter.
The cost per day for 4 rough carpenters.
The cost of 4 rough carpenters for 7 days.

STEP 2. PLAN — Find the cost per day for one rough carpenter, then multiply by the number of carpenters; and then multiply by the number of days.

Copyright © by Glencoe Division

WORKSHOP 19 (continued)

STEP 3. WORK

8 hrs per day × \$18 per hr = \$144 per day for 1 carpenter
4 carpenters × \$144 per day for one carpenter = \$576 per day for 4 carpenters
7 days × \$576 per day for 4 carpenters = \$4032 for 4 carpenters for 7 days

STEP 4. ANSWER

It will cost the remodeler \$4032 for 4 rough carpenters.

1. It takes 5 finish carpenters 6 days to do the work. Each finish carpenter earns \$23.50 per hour and works 7.5 hours per day. How much will it cost for the finish carpenter work? ____________

2. Mary Tulley makes a car payment of \$249.50 per month for four years to pay off a car loan of \$11,125. How much did Mary pay in finance charges for the car loan? ____________

WORKSHOP 20: PROBLEM-SOLVING—IDENTIFYING INFORMATION

Before you begin to solve a word problem, first read the problem through carefully and answer these questions.

A. What are you asked to find?

B. What facts are given?

C. Are enough facts given? Do you need more information than the problem provides?

EXAMPLE Peter Butler is a maintenance programmer for Financial Systems, Inc. He earns \$12.45 per hour and is married and claims 2 withholding allowances. Last week he worked 40 hours at the regular rate and 4 hours overtime at time-and-a-half. He is 28 years old. Find his gross pay last week.

SOLUTION

A. Asked for: Peter Butler's gross pay last week.

B. Facts given: \$12.45 hourly rate
40 hours worked at regular rate
4 hours worked at time-and-a-half

C. Facts needed: none

Gross pay = (\$12.45 × 40) + (\$12.45 × 1.5 × 4)
= \$498 + \$74.70 = \$572.70

1. Patricia Miller purchased a new automobile costing \$18,768.95. After a down payment, she paid for the balance with monthly payments of \$247.95 for five years. How much did Patricia pay in finance charges?

____

2. Last week Corner Mart had gross income of \$54,850.50 with the cost of goods sold being \$48,916.80. This week Corner Mart had gross income of \$58,916.80. How much more in gross income did Corner Mart have this week than last week? ____________

 Copyright © by Glencoe Division

Name ______________________________ Date ______________

REVIEW OF FUNDAMENTALS

WORKSHOP 21: PROBLEM SOLVING—USING MORE THAN ONE OPERATION

Some problems require several operations to solve.

EXAMPLE The Regional Financial Center uses ABC Software Alternatives, Inc. as consultants at a cost of $50 per hour. Over the past 6 months, they have paid ABC $3300. On average, how many hours per month has the Regional Financial Center used ABC Software Alternatives, Inc. over the past six months?

SOLUTION

GIVEN $3300 at $50 per hour over past 6 months

FIND Number of hours per month

WORK Divide $3300 by $50 = 66 hours

Divide 66 hours by 6 months = 11 hours per month

1. The Coffee Cup sold coffee and donuts from a push cart during the last holiday parade. They sold 316 cups of coffee at 75¢ per cup and 21 dozen donuts at 50¢ a donut. What were The Coffee Cup's total sales from its push cart during the last parade? ______________

2. Pioneer Drug Store sold 76 copies of *The Daily Press* at 75¢ a copy. The owner pays 68¢ for each copy of *The Daily Press*. What was the net income from *The Daily Press* for Pioneer Drugs? ______________

WORKSHOP 22: PROBLEM SOLVING—USING ESTIMATION

An important part of problem solving is determining the reasonableness of an answer. Estimation can be used to check the reasonableness of an answer.

EXAMPLE Three cans of tomato sauce cost $1.49, six cans of soda cost $2.39, and six oranges cost $1.79. About how much will it cost for one of each item?

SOLUTION

$1.49: 3 is about $0.50 each

$2.39: 6 is about $0.40 each

$1.79: 6 is about $0.30 each

Total is about: $1.20

1. Mary Sparks purchased 12 gallons of gas at 98.9¢ a gallon. Using her calculator, she computed the cost to be $1186.80. This did not seem reasonable. Is it? What did Mary do wrong?

____

____

2. Mike Moore found a $49.95 sweater on sale at 30% off. He mentally figured the sale price to be about $15. Is he right or wrong? Why?

____

____

3. The Turners found a new home they wish to purchase. The selling price of the house is $118,900. The Turners' financing requires a 20% down payment plus approximately $2500 in closing costs. The Turners estimate they will need about $26,500 to buy this house. Are they right or wrong? Why?

____

Copyright © by Glencoe Division

Name ____________________ Date ____________

REVIEW OF FUNDAMENTALS

WORKSHOP 23: PROBLEM SOLVING—CONSTRUCTING A TABLE

Constructing a table can be a good way of solving some problems. By organizing the data into a table, it is easier to identify the information that you need.

EXAMPLE Bill Calter is an appliance salesperson. For each new refrigerator he sells, he earns 25 bonus points; for each range he sells, he earns 20 bonus points. Bill earned 360 bonus points by selling 16 appliances last week. How many refrigerators and ranges did he sell?

SOLUTION Set up a table to evaluate the possibilities.

Refrigerators $\times$ 25	Ranges $\times$ 20	Total Points	
16	0	400	found by $(16 \times 25) + (0 \times 20)$
15	1	395	found by $(15 \times 25) + (1 \times 20)$
14	2	390	found by $(14 \times 25) + (2 \times 20)$

Note that total points are reduced by 5 each time a refrigerator is switched to a range. To get to 360 total points, 6 more switches are needed: $(390 - 360) \div 5$.

8	8	360	found by $(8 \times 25) + (8 \times 20)$

1. If you had 30 nickels and dimes worth $2.40, how many of each kind of coin do you have?

____

2. The Children's Furniture Shop used 416 legs to assemble a total of 110 3-legged milking stools and 4-legged chairs. How many milking stools were assembled? How many chairs were assembled?

____

WORKSHOP 24: PROBLEM SOLVING—LOOKING FOR A PATTERN

Some problems can be solved more easily if the information is first put into a list or table, then the list or table is examined to see if a a pattern emerges.

EXAMPLE Write the next three numbers for the established pattern.

3, 6, 11, 18, . . .

SOLUTION Notice that 3 was added to 3 resulting in 6
5 was added to 6 resulting in 11
7 was added to 11 resulting in 18
therefore, add $9 + 18 = 27$
add $11 + 27 = 38$
add $13 + 38 = 51$

The next three numbers are 27, 38, and 51.

For problems 1–3, look for a pattern and then write the next three numbers.

1. 1, 5, 9, 13, 17 ____________

2. 1, 2, 5, 14, 41 ____________

3. 32, 16, 8, 4 ____________

4. A one-dozen package of buns costs $1.49 while an eight-pack of buns costs $1.19. If you need 50 buns, what is the least amount you can pay and buy at least 50 buns? ____________

Copyright © by Glencoe Division

Name ______________________________ Date ______________

REVIEW OF FUNDAMENTALS

WORKSHOP 25: PROBLEM SOLVING—GUESS AND CHECK

One way to solve a problem is the guess and check or trial and error method. Guessing at the solution means making an informed guess and then checking it against the conditions stated in the problem to determine how to make a better guess.

EXAMPLE Victor Doss wants to fence off a rectangular garden. He has 80 feet of fencing. What are the dimensions that will give him the largest area?

SOLUTION One length and one width (half-way around) must total 40 feet since the total distance around must total 80 feet. If Victor were to use 20 feet by 20 feet, he would have an area of 400 square feet. Can he do any better than that?

Length	20'	18'	25'	21'	
Width	20'	22'	15'	19'	
Area		400 sq ft	396 sq ft	375 sq ft	399 sq ft

No. Victor should make his garden 20' × 20' to have the largest area.

1. If each different letter stands for a different number, but the same letter always stands for the same number, solve this problem

$$\begin{array}{r} S\ E\ N\ D \\ +\ M\ O\ R\ E \\ \hline M\ O\ N\ E\ Y \end{array}$$

____

2. Batteries come in packs of 3 or 4. If your class needs 20 batteries, how many different ways are there of buying exactly 20 batteries? Which combination do you think would be the cheapest? ______________

WORKSHOP 26: PROBLEM SOLVING—WORKING BACKWARDS

Word problems that involve a sequence of events or actions can sometimes be solved most efficiently by working backwards. If the final result of the problem is given, start your solution at that point and work backwards through the steps to arrive at the beginning conditions of the problem.

EXAMPLE Each year a car is worth 80% of its value the previous year. A car is now worth $10,240. What was its value two years ago?

SOLUTION GIVEN: Car is now worth $10,240. Worth 80% of its value the previous year.

FIND: Value of car two years ago.

WORK: Value of car last year: $10,240 ÷ 80% = $12,800

Value of car two years ago: $12,800 ÷ 80% = $16,000

1. Forty-three players enter the company tennis tournament. How many matches must be played if:

a. it is a single elimination tournament? ______________

b. it is a double elimination tournament and the winner is undefeated? ______________

c. it is a double elimination tournament and the winner has one loss? ______________

2. A recipe for 24 medium-sized pancakes requires two eggs. Eggs are sold in cartons containing 12 eggs. The company breakfast plans to serve 360 people an average of three pancakes each. How many cartons of eggs are needed? ______________

Copyright © by Glencoe Division

Name ______________________ Date ______________

REVIEW OF FUNDAMENTALS

WORKSHOP 27: PROBLEM SOLVING—DRAWING A SKETCH

Some word problems can be simplified if you draw a sketch.

EXAMPLE Pam O'Connor leaves her home and jogs six blocks east, four blocks south, five blocks west, three blocks north, and five blocks west when she stops for a rest. Where is she in relation to her home?

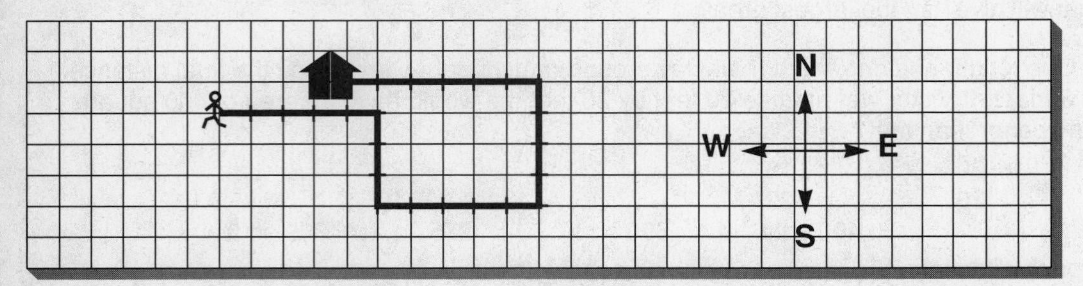

Pam is one block south and four blocks west of her home.

1. A six-volume set of encyclopedias is arranged as shown. If each cover is 0.3 cm thick and the text in each volume is 4 cm thick, what is the distance from page one, volume I, to the last page of volume VI? ______

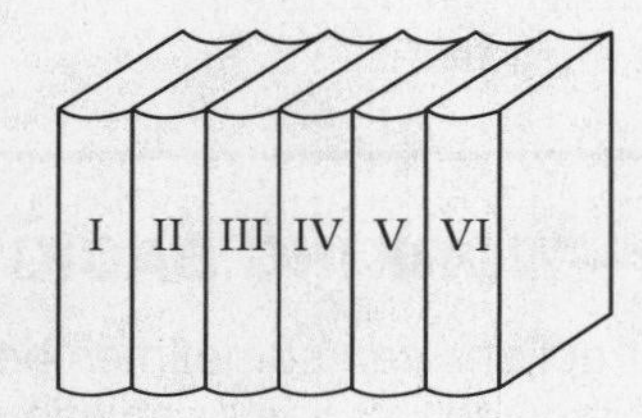

2. A dog is on a 15-foot leash tied to the corner of a 12-foot by 20-foot shed. The dog always stays outside the shed. How many square feet of ground can the dog reach? (Use pi = 3.14) ______

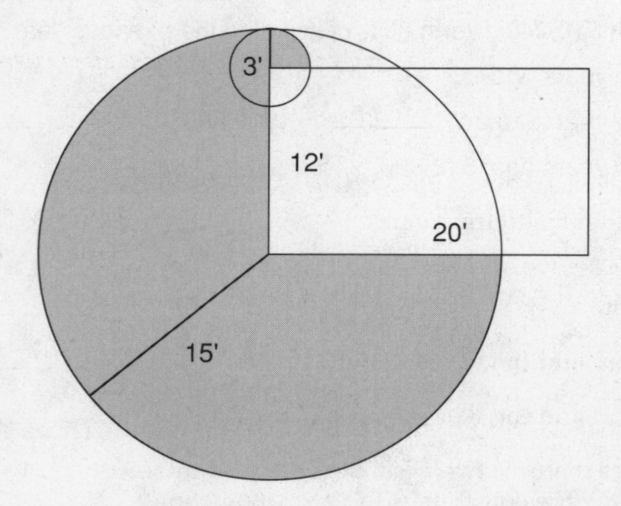

Copyright © by Glencoe Division

Name ______________________ Date ______________

REVIEW OF FUNDAMENTALS

WORKSHOP 28: PROBLEM SOLVING—MAKING A DIAGRAM

Venn Diagrams show relationships between several groups of people or objects and can be used to solve problems.

EXAMPLE Of the 40 employees at Morse Tool and Die, 30 are signed up for health care, 20 are signed up for vision care, 18 are signed up for dental care, 14 are signed up for health and vision care, 13 for health and dental, 6 for vision and dental, and 3 for all three. How many are not signed up for any of the three?

SOLUTION Start with the three signed up for all three. That leaves three for vision and dental, ten for health and dental, and eleven for health and vision. Then that leaves two for dental only, three for vision only, and six for health care only.

The total accounted for is 38, found by $6 + 3 + 2 + 11 + 10 + 3 + 3$ and $40 - 38$ leaves 2 not signed up for anything.

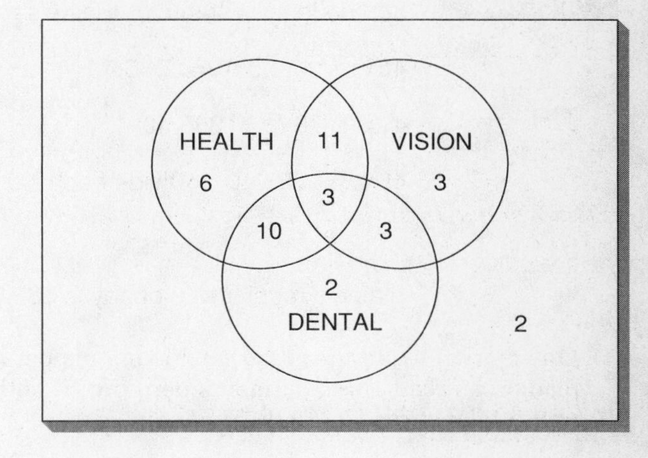

1. A survey of 200 people revealed that 170 owned an answering machine, 125 owned a cordless phone, and 98 owned both. How many of the 200 do not own either?

____

2. Of 20 investors interviewed on the evening news, 13 owned stock, 18 owned mutual funds, 11 owned gold, 11 owned stocks and mutual funds, 10 owned mutual funds and gold, 8 owned stocks and gold, and 7 owned all three. How many investors owned gold but not stocks or mutual funds?

____

Copyright © by Glencoe Division

Name ______________________ Date ______________

REVIEW OF FUNDAMENTALS

WORKSHOP 29: PROBLEM SOLVING—WRITING AN EQUATION

A word problem can frequently be translated into an equation that is solved by performing the same mathematical operation to both sides. Solving the equation then leads to the solution of the problem.

EXAMPLE Last week Elaine Allen worked 40 hours at her regular rate of pay plus 4 hours at time-and-a-half (1.5 times her regular rate of pay). Her gross pay last week was $386.40. What is her regular rate of pay?

SOLUTION Let the letter x stand for Elaine's regular rate of pay, then

$$40x + 4(1.5x) = 386.40$$

$$40x + 6x = 386.40$$

$$46x = 386.40 \quad \text{(Divide each side by 46.)}$$

$$x = 8.4$$

Elaine's regular rate of pay is $8.40 per hour.

1. One optical reader can read 99 documents per hour while a second optical reader can read 105 documents per hour. If both readers are used, how long will it take to read 1326 documents?

____

2. General News stock sells for $32\frac{1}{8}$ a share. The Discount Brokerage House charges a flat fee of $50 per transaction. How many shares of General News stock could you buy for $7760?

____

WORKSHOP 30: PROBLEM SOLVING—RESTATING THE PROBLEM

To solve a word problem you may find it helpful to restate the problem in a different way.

EXAMPLE Your automobile gas mileage is 30 miles per gallon. Your vacation location is 1200 miles from your home. If gas costs an average of $1.09 per gallon, how much will you spend for gas to go and return from your home to your vacation location?

SOLUTION Restate the problem to follow a series of simpler problems.

A. What is the round trip distance? 2×1200 miles $=$ 2400 miles

B. How many gallons of gas would you use? $2400 \div 30 = 80$ gallons

C. How much would the gas cost? $80 \times \$1.09 = \$87.20$

1. A bicyclist travels at an average rate of 14 mph. If she rides 4 hours a day, 6 days a week, how long will it take her to travel 6720 miles? ________

2. There are 22 teams in the National Hockey League. To conduct their annual draft, teams in each city must have a direct telephone line to each of the other teams. How many direct telephone lines must be installed to accomplish this? ________

Copyright © by Glencoe Division

Name ______________________ Date ______________

LESSONS 1-1, 1-2

Hourly Pay and Overtime Pay

Some jobs pay a fixed amount of money for each hour you work. The hourly rate is the amount of money you earn per hour. Straight-time pay is the total amount of money you earn for a pay period at the hourly rate. The overtime rate may be $1\frac{1}{2}$ times your regular hourly rate.

STRAIGHT-TIME PAY = HOURLY RATE × HOURS WORKED

OVERTIME PAY = OVERTIME RATE × OVERTIME HOURS WORKED

TOTAL PAY = STRAIGHT-TIME PAY + OVERTIME PAY

---

1. Ken Douglas is a truck driver. He earns $17.65 per hour. He worked 40 hours last week. What is his straight-time pay? ______

2. Helen Kuhlman is a secretary. She earns $7.65 per hour. She worked 38 hours last week. What is her straight-time pay? ______

3. What is the straight-time pay for each of the following for a 40-hour week?

PRIVATE CLUB SERVICE HELP
Has openings for Locker room attendant (men's) to $5.90 per hr., Bus server to $5.80 per hr., Servers to $6.73 per hr., Pantry help to $5.30 per hr. Send resume to P.O. Box 1073, Toledo, OH

a. locker room attendant ______ **b.** server ______

b. pantry help ______ **d.** bus server ______

4. As an electronics technician, Laura Kale earns $10.92 per hour. She earns double time for work on Sundays. Last week Laura worked 38 regular hours plus 5 hours on Sunday. What was her total pay for the week? ______

5. Eli Katcher is hired as a landscaper. He works $7\frac{3}{4}$ hours each day, Monday through Friday. What would his total pay be for the week if he is paid at the:

LANDSCAPING/WILL TRAIN
$5-$7.90 HOUR
CAREER CONNECTIONS
Free Resume 865-9786 small fee

a. bottom of the pay scale? ______

b. top of the pay scale? ______

6. Georgia Kassem installs fencing. She is paid $5.50 per hour for an 8-hour day and time and a half for overtime for any work over 8 hours per day. What is her pay for a week when she worked 7, 12, 9, 10, and 10 hours? ______

7. Tom Klein is paid $8.20 per hour and time and a half for any work over 40 hours per week. Find his total hours and gross pay for each week.

	M	T	W	Th	F	S
Week of 3/10	5.5	6	9	12	7	0
Week of 3/17	8	8.5	9	13	4.5	4

______ ______

______ ______

8. You are a computer technician for Data Control Company. You earn a regular hourly rate of $10.40. You earn time and a half for overtime work on Saturdays and double time on Sundays. This week you worked 38 hours from Monday through Friday, 8 hours on Saturday, and 5 hours on Sunday. What is your total pay for the week? ______

Copyright © by Glencoe Division

Name ______________________ Date ______________

LESSON 1-3

Weekly Timecards

When you work for a business that pays on an hourly basis, you are usually required to keep a timecard. The timecard shows the time you reported for work and the time you departed each day.

TOTAL HOURS = SUM OF DAILY HOURS

Compute the hours worked for each day on the timecards. Round each day to the nearest quarter hour. What are the total hours for the week? What is the total pay for the week?

1. **TEMPORARY EMPLOYEE TIMECARD**

NAME: Amanda Tackett

DEPT: Accounting

Note: No overtime rate.

Amanda Tackett

EMPLOYEE SIGNATURE

DATE	IN	OUT	IN	OUT	HOURS
9/13	7:00	11:00	11:30	4:45	
9/14	8:15	11:35	12:30	4:35	
9/15	8:10	12:00	12:40	4:10	
9/16	8:20	11:50	12:50	4:50	
9/17	7:05	11:09	11:50	3:30	
RATE PER HOUR: $5.50			TOTAL HOURS		

______

2. **TEMPORARY EMPLOYEE TIMECARD**

NAME: Ervin Vost

DEPT: Auto Parts

Note: Overtime is $1\frac{1}{2}$ on hours over 40.

Ervin Vost

EMPLOYEE SIGNATURE

DATE	IN	OUT	IN	OUT	HOURS
4/5	8:00	12:00	1:00	5:00	
4/6	8:10	11:40	12:40	4:10	
4/7	7:55	12:05	12:55	5:20	
4/8	8:30	11:35	12:15	4:30	
4/9	7:50	11:55	12:40	5:00	
4/10	7:30	11:45			
RATE PER HOUR: $5.40			TOTAL HOURS		

______

3. **TEMPORARY EMPLOYEE TIMECARD**

NAME: Eugene Mueller

DEPT: Sales

Note: No overtime rate.

Eugene Mueller

EMPLOYEE SIGNATURE

DATE	IN	OUT	IN	OUT	HOURS
8/8	7:00	11:15	12:10	4:10	
8/9	6:50	11:00	11:50	4:00	
8/10	7:10	11:46	12:34	3:58	
8/11	7:10	11:10	12:00	4:35	
8/12	7:05	10:55	11:41	3:25	
RATE PER HOUR: $4.50			TOTAL HOURS		

______

4. **TEMPORARY EMPLOYEE TIMECARD**

NAME: Nina Surovy

DEPT: Factory Shop

Note: Overtime is $1\frac{1}{2}$ on hours over 8 and all Saturday hours.

Nina Surovy

EMPLOYEE SIGNATURE

DATE	IN	OUT	IN	OUT	HOURS
3/19	3:30	7:30	8:15	12:10	
3/20	3:25	7:40	8:30	11:25	
3/21	3:32	7:29	8:10	11:30	
3/22	3:23	7:25	8:05	12:00	
3/23	3:40	7:40			
3/24	12:15	6:15			
RATE PER HOUR: $13.28			TOTAL HOURS		

______

Copyright © by Glencoe Division

Name ______________________ Date ______________

LESSON 1-4

Piecework

Some jobs pay on a piecework basis. You are paid for each item of work that you produce.

TOTAL PAY = RATE PER ITEM × NUMBER PRODUCED

1. Linda Hart is a punch press operator for Drummond Machine Company. She earns $0.95 for every molding she presses. What is her total pay for a week in which she presses 310 moldings? ______

2. Frank Holmes delivers newspapers for Farmland News. He is paid 13¢ for every daily paper he delivers and 22¢ for every Sunday paper. What is Frank's pay for a week in which he delivers 574 daily papers and 165 Sunday papers? ______

3. Jeff Lind delivers Danberry Real Estate calendars to town residents after school. He is paid 7.4¢ for each calendar he delivers. What is his total pay for a week in which he delivered the following number of calendars? ______

Mon.	Tues.	Wed.	Thur.	Fri.
100	130	145	138	210

4. Nina Milling assembles transistor radios. She is paid $0.54 for each radio assembled during a regular work week, $0.62 for each radio assembled on Saturdays, and $0.78 for each radio assembled on Sundays. What is Nina's total pay for a week in which she assembled the following number of radios? ______

Mon.	Tues.	Wed.	Thur.	Fri.	Sat.	Sun.
128	117	90	120	122	70	55

5. You are a packer for Acme Bottling Corp. You fill cartons with bottles. Monday through Friday you are paid $0.41 for each carton you fill. On Saturday you are paid an additional $0.21 for each carton you fill. What is your total pay for a week in which you filled the following number of cartons? ______

Mon.	Tues.	Wed.	Thur.	Fri.	Sat.
64	78	80	82	90	70

6. You work in the upholstery department of a furniture factory. You are trained to upholster couches, loveseats, and chairs. You receive $105 for each couch, $85 for each loveseat, and $65 for each chair. During the last 4 weeks you upholstered 5 couches, 6 loveseats, and 5 chairs.

a. What is your total pay for the 4 weeks? ______

b. If you worked 160 hours, what was your hourly rate? ______

Copyright © by Glencoe Division

Name ______________________ Date ______________

LESSON 1-5

Salary

A salary is a fixed amount of money that you earn on a regular basis. Your salary may be paid weekly, biweekly, semimonthly, or monthly. Your annual salary is the total salary you earn during a year.

$$\text{SALARY PER PAY PERIOD} = \frac{\text{ANNUAL SALARY}}{\text{NUMBER OF PAY PERIODS PER YEAR}}$$

---

1. Barry Mulligan is a clerk-typist. His annual salary is $12,410. What is his biweekly salary? ______

2. Bert Miller was just hired as an engineer for the Saxon Engineering Company. His starting salary is $37,600 annually. What is his semimonthly salary? ______

3. Ralph Konsky was hired as a legal consultant for Hereford Cattle Company. His annual salary is $34,100. What is his weekly salary? ______

4. Mary O'Connel is earning a weekly salary of $275 as a payroll clerk. She has accepted a new assignment in the photo processing department. In her new position, she will be paid an annual salary of $15,340. How much more will she earn per week in her new position? ______

5. Bruce Roberts is earning an annual salary of $18,895. He has been offered the job in the ad. How much more would Bruce earn per month if he is paid the minimum? The maximum? ______ ______

SYSTEMS/PROGRAMMER:
$22-28,000. Fee Paid. If you have a degree in Computer Science plus 2 or more years experience with COBOL, send your resume today. Executive Data Processing. P.O. Box 1746.

6. Catherine Lewiston is currently earning an annual salary of $15,080 at Zimmerman Heating Company. She has been offered a job at Carnegie Steel Company at an annual salary of $16,120. How much more would Catherine earn per week at Carnegie Steel than at Zimmerman Heating? ______

7. Debra Arthur earns a weekly salary of $372 at All Sports. Next month she will be promoted from assistant buyer to head buyer. In her new position she will be paid $831.33 semimonthly. How much more per year will Debra earn as a head buyer than as assistant buyer? ______

8. You are a fast-food manager trainee for McDunckles. You are earning a monthly salary of $1100. Next month you will be promoted from trainee to manager. In your new position you will be paid an annual salary of $14,520. How much more per month will you earn as manager than as trainee? ______

9. Consider the 2 ads for an accounting clerk. If you worked 40 hours per week for 50 weeks, how much would each company pay per hour? ______ ______

ACCOUNTING CLERK: $14,500 Fast paced, CRT, type 60wpm, benefits. 866-9281, 5640 Southwyck

ACCOUNTING CLERK: $10,000/yr Salary "Great company" Full benefits! Interviewing now! Call Polly/Sue 243-2222.

Copyright © by Glencoe Division

Name ______________________ Date ______________

LESSONS 1-6, 1-7

Commision and Graduated Commision

A commission is an amount of money that you are paid for selling a product or service. Your commission rate may be a specified amount of money for each sale or it may be a percent of the total value of your sales. A graduated commission offers a different rate of commission for each of several levels of sales.

COMMISSION = COMMISSION RATE × TOTAL SALES

TOTAL GRADUATED COMMISSION = SUM OF COMMISSIONS FOR ALL LEVELS OF SALES

---

1. Ken Oberlin sells cleaning supplies for Soapcraft. He receives a straight commission of 12% of the selling price of each item sold. What commission will he receive for selling $2591.67 worth of cleaning supplies? ________

2. Marion Robarge sells cosmetics. Her commission is 30% of the selling price of every item she sells. What is her commission when she sells cosmetics totaling $717.53? ________

3. You are selling storm windows and doors. You receive a straight commission of 15% of the selling price of each storm window and 20% of the selling price of each door. What commission will you receive for selling storm windows worth $1535 and doors worth $790? ________

4. Audrey Sanders sells sports equipment. She is paid a commission of 5.14% of her first $1500 in sales during the week and 10% on all sales over $1500. What is her commission in a week in which she sells $2510 worth of sports equipment? ________

5. Jasper Carrott sells farm chemicals. He is paid a commission of 9% of his first $6000 in sales during the month and 14% on all sales over $6000. What is his commission in a month in which he has sales worth $16,120? ________

6. Dan Valdez sells appliances for McKenny Appliance Store. He is paid an 8% commission on the first $1000 of sales, 10% on the next $1500, and 15% on all sales over $2500. What is his commission on $4910 in sales? ________

7. Elanor Fray demonstrates microwave ovens at the National Home Show. She is paid $7 each for the first 7 demonstrations and $11 for each demonstration over 7. What is Maria's commission for a day in which she makes 12 demonstrations? ________

8. You are paid a commission plus $3.40 per hour with time-and-a-half overtime for all hours over 8 per day. Your commission consists of 4% of the first $3000 in sales and 5% on all sales over $3000. Find your gross pay for a week in which you worked 9 hours on Monday, 8 hours on Tuesday, 11 hours on Wednesday, 9 hours on Thursday, 10 hours on Friday, and 8 hours on Saturday. Your total sales for the week were $4115. ________

9. You are the sales manager for the Shockey Solar Panel Company. You receive a commission based on the total sales of all the salespeople you manage. Your commission is 2% of the first $80,000 of sales, 4% of the next $80,000, and 5% of all sales over $160,000. What is your commission for a month in which your salespeople sold $242,100 in solar panels? ________

Copyright © by Glencoe Division

Name ____________________ Date ____________

LESSON 2-1

Federal Income Tax

Employers are required by law to withhold a certain amount of your pay for federal income tax (FIT). The Internal Revenue Service provides employers with tables that show how much money to withhold. The amount withheld depends on your income, marital status, and withholding allowances. You may claim one allowance for yourself and one for your spouse if you are married. You may claim additional allowances for any others you support.

Use the tables below to find the amount withheld.

WEEKLY PAYROLL SINGLE					
Wages		Allowances			
At least	But Less than	0	1	2	3
195	200	26	20	14	7
200	210	27	21	15	9
210	220	29	22	16	10
220	230	30	24	18	12
230	240	32	25	19	13
240	250	33	27	21	15
250	260	35	28	22	16
260	270	36	30	24	18
270	280	38	31	25	19
280	290	39	33	27	21

WEEKLY PAYROLL MARRIED					
Wages		Allowances			
At least	But Less than	0	1	2	3
190	195	19	12	6	0
195	200	19	13	7	1
200	210	21	14	8	2
210	220	22	16	10	3
220	230	24	17	11	5
230	240	25	19	13	6
240	250	27	20	14	8
250	260	28	22	16	9
260	270	30	23	17	11
270	280	31	25	19	12

1. Beverly Hibbs earns $224.46 a week. She is single and claims 2 allowances. What amount is withheld weekly for federal income tax? ____________

2. Kent Beals earns $274.10 a week. He is married and claims 3 allowances. What amount is withheld weekly for federal income tax? ____________

3. Pam Edie earns $255.00 per week as a manager at Marlin Department Store. She is married and claimed 1 allowance last year. She hopes to receive a refund on her next tax return by claiming no allowances this year. How much more in withholdings will be deducted weekly if she claims no allowances? ____________

4. Andrew Kendall's gross pay for the week is $251.50. He is married and claims 2 allowances. Starting next week he will receive $20 more per week. How much more per week will he pay in federal income taxes? ____________

5. Roxanne Ray earns $210.50 per week as a sales assistant. Next week she will be promoted to assistant manager. She will then earn $254.00 per week. She is single and claims 1 allowance. How much more will she pay in federal income tax? ____________

6. You are an apprentice plumber for Pointer Plumbing. You are earning an annual salary of $13,936. You are married and claim 3 allowances. What amount is withheld from your weekly pay for federal income tax? ____________

7. You are single and claim 1 allowance. You presently earn $221 per week. Starting next week you will receive a 5% increase in pay and will earn $232.05. How much more will you have withheld from your weekly pay for federal income tax purposes? ____________

8. Nico Joven presently earns $206.80 per week. He claims himself and his mother as allowances. He is single. He is to get a new job classification which will result in a 15% increase in pay. He also plans to marry. After his marriage, he will claim his wife, himself, and his mother as allowances. How much less will be withheld weekly for federal income tax purposes? ____________

Copyright © by Glencoe Division

Name ______________________________ Date ______________

LESSONS 2-2, 2-3

State Income Tax and Graduated State Income Tax

Most states require employers to withhold a certain amount of your pay for state income tax. In some states, the tax withheld is a percent of your taxable wages. Your taxable wages depend on personal exemptions allowed for supporting yourself and others in your family. Most states have a graduated income tax that involves a different tax rate for each of several levels of income.

TAXABLE WAGES = ANNUAL GROSS PAY − PERSONAL EXEMPTIONS

ANNUAL TAX WITHHELD = TAX RATE × TAXABLE WAGES

$$\text{TAX WITHHELD PER PAY PERIOD} = \frac{\text{ANNUAL TAX WITHHELD}}{\text{NUMBER OF PAY PERIODS PER YEAR}}$$

For problems 1-4, use the personal exemptions below and find the amount withheld.

PERSONAL EXEMPTIONS: Single — $1400; Married — $2800;
Each Dependent — $1000

1. Jack Pasler earns $20,940 annually as an accountant. He is married and supports 1 child. The state tax rate in Jack's state is 4.5% of taxable income. What amount is withheld yearly for state income tax? ______________

2. Rose Klaus earns $34,850 annually as an engineer. She is single and supports her father. The state tax rate in Rose's state is 3.0% of taxable income. What amount is withheld yearly for state income tax? ______________

3. Wes Kennedy earns $30,140 annually as a new staff attorney. He is single. The state tax rate in Wes's state is 3.5% of taxable income. What amount is withheld from Wes's monthly pay for state income tax? ______________

4. Katrina Harold earns $364.85 per week. She is married and claims 3 children as dependents. The state tax rate in Katrina's state is 2.0% of taxable income. What amount is withheld from Katrina's weekly pay for state income tax? ______________

5. You are a computer operator for Data Base, Inc. The state has the following personal exemptions and graduated income tax.

Personal Exemptions	
Single	$1500
Married	$3000
Each Dependent	$ 700

STATE INCOME TAX	
Annual Gross Pay	**Tax Rate**
First $2000	1.5%
Next $3000	2.0%
Next $3000	2.5%
Over $8000	3.0%

You earn $20,160 a year. You are single and are paid on a weekly basis. What amount is withheld from your weekly paycheck for state income tax? ______________

Copyright © by Glencoe Division

Name ____________________ Date ____________

LESSON 2-4

Social Security and Medicare Tax

The Federal Insurance Contributions Act (F.I.C.A.) requires employers to deduct 6.2% of the first $62,700 of your annual income for social security taxes and 1.45% of all your annual income for medicare taxes. The employer must contribute an amount that equals your contribution.

TAX WITHHELD = TAX RATE × GROSS PAY

Use the social security tax rate of 6.2% of the first $62,700 and medicare tax rate of 1.45% of all annual income to solve.

1. Roland Purcell, a draftsman, has a gross monthly income of $1900. His earnings to date for this year are $20,900. How much is deducted this month for social security? How much for medicare? ________ ________

2. Mark Traynor, a store manager, has a gross weekly income of $412.96. His earnings to date for this year total $19,822.08. How much is deducted from his paycheck this week for social security? How much for medicare? ________ ________

3. Maria Whetstone, a boiler operator, has a gross weekly income of $502.16. Her earnings to date for this year are $20,086.40. How much is deducted from her paycheck this week for social security? How much for medicare? ________ ________

4. Tom Zetts is an aerospace engineer. He earns $61,710 a year and is paid on a semimonthly basis. How much is deducted per pay period for social security? How much for medicare? ________ ________

5. Renae Walters is paid a salary of $6000 per month.

a. How much is deducted in January for social security? For medicare? ________ ________

b. How much is deducted in December for social security? For medicare? ________ ________

6. Craig Linn earns $75,000 per year. He is paid monthly.

a. How much is deducted in October for social security? For medicare? ________ ________

b. How much is deducted in November for social security? For medicare? ________ ________

c. How much is deducted in December for social security? For medicare? ________ ________

In a few years the social security tax rate will be 6.2% of the first $64,200.

7. a. Rework Problem 5 using this information. ________ ________

________ ________

b. Rework Problem 6 using this information. ________ ________

________ ________

________ ________

8. You earn $1250 per week. How many more paychecks would you have social security withheld at the new maximum of $64,200? How much more per year would be withheld? ________ ________

Name ________________________________ Date ______________

LESSON 2-5

Group Insurance

Many businesses offer group insurance plans to their employees. You can purchase group insurance for a lower cost than individual insurance. Businesses often pay part of the cost of the insurance. The remaining cost is deducted from your pay.

$$\text{DEDUCTION PER PAY PERIOD} = \frac{\text{TOTAL AMOUNT PAID BY EMPLOYEE}}{\text{NUMBER OF PAY PERIODS PER YEAR}}$$

---

1. Ruth Lockwood earns $1680 a month as a supervisor for Acme Steel Company. Her group medical coverage costs $1750 a year. The company pays 75% of the cost of group insurance. How much is deducted each month from Ruth's paycheck for medical insurance? ________

2. David Herbert earns $1460 a month as a security guard for Baldwin Enterprises, Inc. His group medical insurance costs $2110 a year. The company pays 80% of the cost of group insurance. How much is deducted each month from David's paycheck for medical insurance? ________

3. Kasey Marsenburg is a financial consultant for ABC Finance. She earns $456.21 weekly. Her annual group medical coverage costs $1942, of which ABC Finance pays 65%. How much is deducted weekly from Kasey's paycheck for medical coverage? ________

4. Dale Lawrence earns $478 a week as a fireman for the city. The city pays 85% of the cost of any insurance coverage. His family medical insurance costs $1575 a year. How much is deducted each week for medical insurance? ________

5. Harry Marsh earns $365 per week as a chef. He has group medical and dental insurance. Medical coverage costs $1822 a year and dental coverage costs $206 a year. The restaurant pays 80% of the medical and 65% of the dental insurance. How much is deducted each week for both insurance coverages? ________

6. Judy Carol has medical, prescription drug, and term life insurance coverages through the company for which she works. Medical coverage costs $2250 a year and the company pays 80%. Prescription drug coverage costs $192 a year and the company pays 60%. Term life insurance is paid entirely by the company. How much is deducted each week for the coverage? ________

7. You are a traveling sales representative for Lawn-Care, Inc. You have medical, dental, term life insurance, and travel insurance coverages through the company. Medical coverage costs $1580 a year, dental coverage costs $194 a year and travel insurance costs $92 a year. The company pays 80% of the medical, 70% of the dental, and 85% of the travel insurance. Term life insurance is entirely paid for by the company. How much is deducted each month from your paycheck for these coverages? ________

8. You teach in the local public schools. You are married and have family coverage for medical, dental, vision, prescription, and term life insurance. The annual premiums are: medical insurance, $2690; dental insurance, $268; vision insurance, $120; prescription drug coverage, $194; and term life insurance, $270. The Board of Education pays 85% of the medical and prescription insurance, 80% of the dental and vision insurance, and 90% of the term life insurance. You get paid semimonthly from September through May. How much is deducted from each paycheck to pay for these coverages? ________

Copyright © by Glencoe Division

Name ______________________________ Date ______________

LESSON 2-6

Earnings Statement

You may have additional deductions taken from your gross pay for union dues, contributions to community funds, savings plans, and so on. The earnings statement attached to your paycheck lists all your deductions, your gross pay, and your net pay for the pay period.

NET PAY = GROSS PAY − TOTAL DEDUCTIONS

---

1. Find the net pay.

DEPT.	EMPLOYEE	CHECK #	WEEK ENDING	GROSS PAY	NET PAY
236	Pinto, J.	54316	11/5/-	$452.50	

TAX DEDUCTIONS					PERSONAL DEDUCTIONS		
FIT	SOC. SEC.	MED.	STATE	LOCAL	MEDICAL	UNION DUES	OTHERS
$70.00	$28.06	$6.56	$11.27	—	$15.50	—	$24.20

Complete the earnings statements. The social security tax rate is 6.2% of the first $62,700 and the medicare tax rate is 1.45% of all income.

2. Lucy Kreb is a bookkeeper. Her state personal exemptions are $84.60 a week. State tax rate is 3.5% of taxable income. Local tax is 1.5% of gross pay.

DEPT.	EMPLOYEE	CHECK #	WEEK ENDING	GROSS PAY	NET PAY
25	Kreb, L.	3074	3/7/-	$244.00	

TAX DEDUCTIONS					PERSONAL DEDUCTIONS		
FIT	SOC. SEC.	MED.	STATE	LOCAL	MEDICAL	UNION DUES	OTHERS
$27.00					$9.15	$3.00	$3.50

3. Melissa Gregory is an installer for General Telephone. Her state personal exemptions total $98.00 per week. The state tax rate is 3.0% of taxable income. Medical insurance costs $1940 a year, of which the company pays 60% of the costs.

DEPT.	EMPLOYEE	CHECK #	WEEK ENDING	GROSS PAY	NET PAY
A	Gregory, M.	8002	6/10/-	$440.00	

TAX DEDUCTIONS					PERSONAL DEDUCTIONS		
FIT	SOC. SEC.	MED.	STATE	LOCAL	MEDICAL	UNION DUES	OTHERS
$57.00				—		—	$28.00

4. Your annual salary is $28,460. Your state personal deductions total $88.00 for a weekly pay period. The state tax rate is 3.6% of taxable income. The city tax rate is 1.7% of gross pay. Medical and group life insurance are paid by the company. Social security and medicare are withheld. Union dues are $156 per year. FIT is $90 per pay period. What is your net pay for each weekly pay period? ______________

5. You are a locksmith and earn $18.80 an hour for a regular 40-hour week with time and a half for overtime. Your state personal deductions total $48.85 a week. The state tax rate is 4.5% of taxable income. The medical insurance costs $1560 a year, of which your employer pays 75% of the costs. Term life insurance costs $195 a year, of which the company pays 50%. Social security and medicare are withheld. What is your net pay for a week in which you worked 44 hours, if your federal income tax withholding is $127.00? ______________

Name ______________________ Date ______________

LESSON 3-1

Deposits

A deposit is an amount of money that you put into a bank account. You use a deposit slip to record the amounts of currency, coins, and checks that you deposit. To open a checking account, you must make a deposit.

TOTAL DEPOSIT = (CURRENCY + COINS + CHECKS) − CASH RECEIVED

Find the subtotal and the total deposit.

1.

		DOLLARS	CENTS
Cash	Currency		
	Coins		
Checks	LIST SEPARATELY		
	117-4	594	44
	71-97	301	03
	SUBTOTAL		
	➧ LESS CASH RECEIVED	40	00
	TOTAL DEPOSIT		

2.

		DOLLARS	CENTS
Cash	Currency	68	00
	Coins	42	95
Checks	LIST SEPARATELY		
	712-08	129	44
	SUBTOTAL		
	➧ LESS CASH RECEIVED		
	TOTAL DEPOSIT		

3. Margo Xavier has a paycheck for $375.42 and a refund check for $24.95. She would like to receive $45 in cash and deposit the remaining amount. What is Margo's total deposit? ____________

4. Joseph Ross deposited a check for $299.45 and a check for $229.52. He received $42 in cash. What was his total deposit? ____________

5. Paul and Ann Sherwin deposited their paychecks for $375.45 and $614.20 and a check from their insurance company for $187.60. They received $500 in cash. What was their total deposit? ____________

6. Morgan Okonski deposits the following in her checking account: 6 five-dollar bills, 5 two-dollar bills, 12 one-dollar bills, 9 half dollars, 6 quarters, 42 dimes, 10 nickels, 15 pennies, and a check for $97.23. What is Morgan's total deposit? ____________

For problems 7 and 8, use the deposit slips below to show your total deposit.

7. Randy Houck deposited his paycheck for $385.15, a refund check from a mail order purchase for $125.95, and $37.20 in cash. What was Randy's total deposit?

8. You have a check for $223.47 and a check for $24.75. You would like to deposit the checks and receive 2 ten-dollar bills, 3 one-dollar bills, 10 quarters, and 10 dimes. What is your total deposit?

7.

		DOLLARS	CENTS
Cash	Currency		
	Coins		
Checks	LIST SEPARATELY		
	SUBTOTAL		
	➧ LESS CASH RECEIVED		
	TOTAL DEPOSIT		

8.

		DOLLARS	CENTS
Cash	Currency		
	Coins		
Checks	LIST SEPARATELY		
	SUBTOTAL		
	➧ LESS CASH RECEIVED		
	TOTAL DEPOSIT		

Copyright © by Glencoe Division

Name ______________________ Date ____________

LESSON 3-2

Writing Checks

After you have opened a checking account and made a deposit, you may write checks. A check directs the bank to deduct money from your checking account to make a payment. Your account must contain as much money as the amount of the check you are writing so as to avoid overdrawing your account.

Write each amount in words as it would appear on a check.

1. $10.91

2. $228.00

3. $57.81

4. $7.62

5. $4030.00

6. $8461.00

7. Complete check #317 using today's date. Make it payable to Hartman Dental Clinic for $86.97 for an office visit.

317

MARIE SERVOSS ________ 19 __

Pay to the Order of ____________________ $ ______

____________________ DOLLARS

BANK OF CAMDEN

memo ________ Marie Servoss

8. Complete check #128 using today's date. Make it payable to First National Bank for $84.02 for car payment #14.

128

WILLIAM DETWILER ________ 19 __

Pay to the Order of ____________________ $ ______

____________________ DOLLARS

FIRST NATIONAL BANK

memo ________ William Detwiler

9. Complete check #142 using today's date. Make it payable to Rural Farm Electric for $364.90 for your monthly bill. Sign your name.

142

________ 19 __

Pay to the Order of ____________________ $ ______

____________________ DOLLARS

THE MERCHANTS BANK

memo ________ ________

Copyright © by Glencoe Division

Name ______________________________ Date ______________

LESSON 3-3

Check Registers

You use a check register to keep a record of the deposits you make and the checks that you write. The balance is the amount of money in your account. You add deposits to the balance. When you write a check, you subtract the amount of the check from the balance.

NEW BALANCE = PREVIOUS BALANCE − CHECK AMOUNT

NEW BALANCE = PREVIOUS BALANCE + DEPOSIT AMOUNT

1. Sherry Smith opened a new checking account by depositing her paycheck for $347.95. The check register shows her transactions since opening her account. What should her new balance be?

CHECK NO.	DATE	CHECKS ISSUED TO OR DESCRIPTION OF DEPOSIT	AMOUNT OF CHECK		AMOUNT OF DEPOSIT		BALANCE	
		BALANCE BROUGHT FORWARD ➧						
101	1/7	E&H Auto Clinic	102	14				
102	1/11	Heising Builders	67	70				

2. Your checkbook balance was $206.42 on March 3. The check register shows your transactions since. What should your new balance be?

CHECK NO.	DATE	CHECKS ISSUED TO OR DESCRIPTION OF DEPOSIT	AMOUNT OF CHECK		AMOUNT OF DEPOSIT		BALANCE	
		BALANCE BROUGHT FORWARD ➧						
284	3/9	Osborne Pharmacy	32	50				
	3/15	Deposit			120	49		
285	3/20	General Telephone	19	80				
286	3/21	Continental Foods	68	82				

3. Your checkbook balance was $492.16 on September 3. Use the check register below to record the following transactions: On 9/5 check #442 for $102.06 payable to Lenny's Deli; on 9/6 check #443 for $228.00 payable to Merchant's Bank; on 9/10 a deposit of $350.00; on 9/12 check #444 for $35.79 payable to Home Pharmacy; on 9/15 check #445 for $42.22 payable to Trenton Shoe Store; on 9/15 check #446 for $72.60 payable to Home Gas Company.

CHECK NO.	DATE	CHECKS ISSUED TO OR DESCRIPTION OF DEPOSIT	AMOUNT OF CHECK		AMOUNT OF DEPOSIT		BALANCE	
		BALANCE BROUGHT FORWARD ➧						

Copyright © by Glencoe Division

Name ______________________________ Date ______________

LESSON 3-4

Bank Statements

When you have a checking account, you receive a statement and canceled checks from the bank each month. Canceled checks are the checks that the bank has paid by deducting money from your account. Your statement lists all your checks that the bank has paid and your deposits that the bank has recorded since your last statement. The statement may include a service charge for handling the account.

PRESENT BALANCE = PREVIOUS BALANCE + DEPOSITS RECORDED − CHECKS PAID − SERVICE CHARGE

	1.	2.	3.	4.	5.
Previous Balance	$ 42.30	$220.72	$ 37.09	$849.83	$ 19.84
Total Deposits	502.00	304.50	806.51	795.14	1136.22
Total Checks	80.60	398.63	329.77	900.62	1108.00
Service Charge	6.75	2.80	2.85	0	48.06
Present Balance					

6. A portion of Sara Pearson's bank statement is shown. Her previous balance was $286.91. What is her present balance?

CHECKS AND OTHER CHARGES			DEPOSITS AND CREDIT		BALANCE
Date	**Number**	**Amount**	**Date**	**Amount**	
9/24	210	$ 39.95	9/10	$164.00	
9/27	211	110.27	9/18	148.90	
9/29	212	97.58	9/30	45.21	
Service charge		2.25			

7. A portion of your bank statement is shown. Your previous balance was $228.73. What is your present balance?

CHECKS AND OTHER CHARGES			DEPOSITS AND CREDIT		BALANCE
Date	**Number**	**Amount**	**Date**	**Amount**	
10/13	142	$ 13.60	10/5	$146.90	
10/18	143	112.76	10/12	188.42	
10/27	144	10.64	10/19	135.65	
10/31	145	38.81			
Service charge		5.14			

Copyright © by Glencoe Division

Name ________________________________ Date ______________

LESSON 3-5

Reconciling the Bank Statement

When you receive your bank statement, you compare the canceled checks, the bank statement, and your check register to be sure they agree. You may find some outstanding checks and deposits that appear in your register but did not reach the bank in time to be processed and listed on your statement. You reconcile the statement to make sure that it agrees with your check register.

ADJUSTED BALANCE = STATEMENT BALANCE − OUTSTANDING CHECKS + OUTSTANDING DEPOSITS

Complete the table.

	1.	2.	3.	4.	5.
Check Register Balance	$275.14	$378.95	$1591.40	$1202.91	$1861.20
Service Charge	4.81	4.35	6.20	0	0
NEW BALANCE					
Statement Balance	549.95	231.36	1027.33	2174.00	2361.40
Outstanding Checks	529.63	190.00	0	1046.20	812.14
Outstanding Deposits	250.01	333.24	557.87	75.11	311.94
ADJUSTED BALANCE					

Do the register and statement balances agree?

6. After comparing your bank statement, canceled checks, and checkbook register, you complete the reconciliation statement shown. What are the new and adjusted balances?

RECONCILIATION STATEMENT

Check Register Balance	$ 285.14	**Statement Balance**	$ 182.63
Service Charge	− 8.10	**Outstanding Checks**	
		#202 $35.92	
		#203 $28.75	
NEW BALANCE	$ ______	______	−$ ______
			$ ______
		Outstanding Deposits	
		$129.08	
		$30.00	+$ ______
****RECONCILIATION STATEMENT For Your Convenience**		**ADJUSTED BALANCE**	$ ______

Do the register and statement balances agree? ______

Copyright © by Glencoe Division

Name ______________________________ Date ______________

Reconciling a N.O.W. Account: A Simulation

Your Negotiable Order of Withdrawal (N.O.W.) Account is similar to a checking account except that it earns interest for you. Interest is the amount you earn for permitting the bank to use your money. You also make your mortgage payment to Peoples Savings through an automatic transfer arrangement. In addition, the interest on your certificate of deposit is automatically credited to your N.O.W. account. Using your N.O.W. account statement from the bank and your checkbook register, complete the reconciliation form on the next page.

N.O.W. ACCOUNT STATEMENT

Check Number	Amount	Date	Description	Bank Reference Number	Date	Balance
	50.00	12/03	Cash Withdrawal: 24-hour	681765	12/01	768.70
315	275.80	12/07	Check	249247	12/03	718.70
	132.42	12/07	CD Interest Credit	756358	12/07	575.32
316	99.22	12/10	Check	249861	12/10	992.30
	516.20	12/10	Deposit	332196	12/11	892.30
	100.00	12/11	Cash Withdrawal: 24-Hour	681983	12/15	579.37
317	312.93	12/15	Check	249187	12/17	435.77
318*	143.60	12/17	Check	249193	12/23	951.97
	516.20	12/23	Deposit	332552	12/28	390.78
322	96.19	12/28	Check	249337	12/31	395.53
	465.00	12/28	Peoples Savings Transaction	817130		
	4.75	12/31	N.O.W. Account Interest	527168		

Items Enclosed	Balance Last Statement	Total Amount Charged	Total Amount Credited	Balance This Statement
5	768.70	1542.74	1169.57	395.53

*Indicates the next sequentially numbered check or checks may have (1) been voided by you, (2) not yet been presented to the bank, or (3) appeared on a previous statement.

PLEASE BE SURE TO DEDUCT ANY PER CHECK CHARGES OR SERVICE CHARGES THAT MAY APPLY TO YOUR ACCOUNT

CHECK OR TRANS. NO.	DATE	DESCRIPTION OF TRANSACTION	CHECK OR TRANSACTION AMT.		√ T	(−) CHECK FEE (IF ANY)	(+) AMOUNT OF DEPOSIT		BALANCE	
									768	70
	12/03	Cash Withdrawal	50	00					718	70
315	12/07	Security National Car Pymt	275	80					442	40
316	12/10	King Food Locker	99	22					343	68
	12/10	Deposit-Paycheck					516	20	859	88
	12/11	Cash Withdrawal	100	00					759	88
317	12/14	Universal Appliance Co.	312	93					446	95
318	12/17	Metro Electric Co.	143	60					303	35
	12/23	Deposit-Paycheck					516	20	819	55
319	12/25	King Drugs	17	93					801	62
~~320~~		VOID								
321	12/27	Suburban News	15	80					785	82
322	12/28	King Food Lockers	96	19					689	63

Copyright © by Glencoe Division

Name ______________________ Date ____________

Reconciling a N.O.W. Account: A Simulation (CONTINUED)

BALANCING MADE EASY

To adjust your checkbook register balance for reconciliation with your N.O.W. Account Statement, enter your checkbook balance here. ____________

1. Add to your checkbook balance all deposits, interest, and other credits posted on this statement that you have not already recorded. **1.** ____________ (+)
2. Subtract all automatic and other charges such as automatic savings posted on this statement that you have not already recorded. **2.** ____________ (−)
3. Subtract from your checkbook balance any Service Charge shown on your statement. **3.** ____________ (−)
4. The steps above will give you your adjusted balance. Enter this adjusted balance here. **4.** [ ]

Now, to reconcile your N.O.W. Account Statement with your adjusted balance in your checkbook register:

1. Write the amount shown on the N.O.W. Account Statement under "Balance this Statement." **1.** ____________
2. Add any deposits you have made that the Bank has not shown.

2. Date	Amount
	$
SubTotal	$ (+)

3. Subtract outstanding checks not shown.

3. Check #	Amount
	$
SubTotal	$ (−)

4. Now you should have the same figure as your adjusted checkbook register. **4.** [ ]

Copyright © by Glencoe Division

Name ______________________________ Date ______________

Deposits and Withdrawals

To open a savings account, you must make a deposit. Each time you make a deposit, it is added to the balance of your account. A withdrawal is an amount of money that you take out of your savings account. When you make a withdrawal, it is subtracted from the balance of your account. Your bank may provide deposit and withdrawal slips for you to fill out for each transaction.

TOTAL DEPOSIT = (CURRENCY + COINS + CHECKS) − CASH RECEIVED

1. Your account number is 718512. You wish to deposit $34 in currency, $6.25 in coins, and a check for $228.91. Complete the savings deposit slip.

COMMUNITY BANK

DEPOSIT SLIP

DATE Today's date 19__

DEPOSIT TO THE ACCOUNT OF:

NAME Your name

PRINT NUMBER | | | | | | |

0100 0059

		DOLLARS	CENTS
CASH	CURRENCY		
	COIN		
CHECKS	LIST SINGLY		
	TOTAL		

2. Your account number is 421746. You wish to deposit 120 dimes, 25 quarters, and checks for $184.63 and $196.17. You wish to receive $50 in cash. Complete the savings deposit slip.

COMMUNITY BANK

DATE Today's date 19__

DEPOSIT TO THE ACCOUNT OF:

NAME Your name

PRINT NUMBER | | | | | | |

0100 0059 DEPOSIT SLIP

		DOLLARS	CENTS
Cash	Currency		
	Coin		
Checks	LIST SEPARATELY		
	SUBTOTAL		
	LESS CASH RECEIVED		
	TOTAL DEPOSIT		

3. Your account number is 0651831. You wish to withdraw $332.16. Complete the savings withdrawal slip.

COMMUNITY BANK — WITHDRAWAL

DATE ______________ SAVINGS ACCOUNT NUMBER | | | | | | | |

______________ (AMOUNT IN WORDS) DOLLARS | dollars | cents |

0100 0059 ______________ SIGN HERE

4. Your account number is 9841730. You wish to withdraw one hundred twenty-three dollars and sixty-four cents. Complete the savings withdrawal slip.

COMMUNITY BANK — WITHDRAWAL

DATE ______________ SAVINGS ACCOUNT NUMBER | | | | | | | |

______________ (AMOUNT IN WORDS) DOLLARS | dollars | cents |

0100 0059 ______________ SIGN HERE

Copyright © by Glencoe Division

Name ______________________________ Date ______________

LESSONS 4-3, 4-4

Passbooks and Account Statements

Your bank may provide you with a savings account passbook. When you make a deposit or withdrawal, a bank teller records in your passbook the transaction, any interest earned, and the new balance. Your bank may mail you a monthly or quarterly account statement showing all deposits, withdrawals, and interest credited to your account since the last statement date.

NEW BALANCE = PREVIOUS BALANCE + INTEREST + DEPOSITS − WITHDRAWALS

	1.	2.	3.	4.	5.	6.
Previous balance	$504.23	$597.70	$882.85	$8412.56	$410.88	$2218.47
Interest	$ 18.72	$ 18.02	$ 40.70	$ 102.82	$ 10.98	$ 72.18
Deposit	$ 98.25	—	—	—	$ 75.00	$ 612.89
Withdrawal	$150.00	—	$ 20.00	$8000.00	$217.49	$ 750.00
New balance						

7. Find the balance for each date on this savings account passbook.

PASSBOOK ACCOUNT NUMBER 07-62345				
DATE	**DEPOSIT**	**WITHDRAWAL**	**INTEREST**	**BALANCE**
1/04				285.85
3/15	25.00			
4/03			8.26	
4/14		175.00		
6/20	814.50			
7/03			15.62	
7/20	814.50			
9/23		125.00		
10/03			19.42	

8. You receive this savings account statement. What is the balance in your account on June 30?

ACCOUNT NUMBER: 01-217843				
DATE	**DEPOSIT**	**WITHDRAWAL**	**INTEREST**	**BALANCE**
4/08		315.28		152.07
5/25	261.30			
6/10	201.20			
6/30			8.75	

PREVIOUS STATEMENT		THIS STATEMENT	
DATE	**BALANCE**	**DATE**	**BALANCE**
3/31	$467.35	6/30	

Copyright © by Glencoe Division

Name ______________________ Date ______________

LESSON 4-5

Simple Interest

When you deposit money in a savings account, you are permitting the bank to use the money. The amount you earn for permitting the bank to use your money is called interest. The principal is the amount of money earning interest. The annual interest rate is the percent of the principal that you earn as interest based on one year.

INTEREST = PRINCIPAL × RATE × TIME

1. Alice Tsongas deposited $600 in a new savings account at Bradinton Savings and Loan Association. No other deposits or withdrawals were made. After 3 months the interest was computed at an annual interest rate of $7\frac{1}{2}$%. How much simple interest did Alice's money earn? ______

2. Henry Bonnacio deposited $1000 in a new savings account at First National Bank. He made no other deposits or withdrawals. After 6 months the interest was computed at an annual rate $6\frac{1}{2}$%. How much simple interest did Henry's money earn? ______

3. On June 1, Elena Moore deposited $610 in a savings account at Metro Savings and Loan Association. At the end of November her interest was computed at an annual interest rate of 6.5%. How much simple interest did Elena's money earn? ______

4. Wu Chen deposited $819.41 in a savings account at the Suburban Trust Company on May 1. At the end of July his interest was computed at an annual rate of 6%. How much simple interest did Wu's money earn? ______

5. On April 1, Pete Weber deposited a refund check for $368.94 in a savings account at Northern Savings and Loan Association. At the end of December the interest was computed at an annual rate of $6\frac{3}{4}$%. How much simple interest did Pete's money earn? ______

6. On March 31, you opened a savings account at Main Street Savings Bank with a deposit of $817.25. At the end of October the interest was computed at an annual rate $5\frac{3}{4}$% and added to the balance in your account.

   a. How much simple interest did your money earn? ______

   b. What was your new balance? ______

7. On February 1, the balance in your account is $516.81. On July 1, you deposit $310.90. Your bank pays $6\frac{1}{4}$% interest.

   a. How much interest have you earned on July 1? ______

   b. What is your balance (including your deposit) on July 1? ______

   c. How much interest have your earned on November 1? ______

   d. What is your balance on November 1? ______

8. On July 1 of last year, your balance was $229.46. On September 1, you deposited $200. On February 1 of this year, you deposited $312.50. Your bank pays $7\frac{1}{4}$% interest. What is the balance in your account on July 1 of this year? ______

Copyright © by Glencoe Division

Name ______________________________ Date ______________

LESSON 4-6

Compound Interest

Interest that you earn in a savings account during an interest period is added to your account. The new balance is used to calculate the interest for the next interest period. Compound interest is cinterest earned not only on the original principal but also on the interest earned during previous interest periods.

AMOUNT = PRINCIPAL + INTEREST

---

1. Michael Arthur deposited $2900 in a new regular savings account that earns 5.5% interest compounded semiannually. He made no other deposits or withdrawals. What was the amount in the account at the end of 1 year? ______________

2. Teresa Marvin deposited $1950 in a new credit union savings account on the first of a quarter. The principal earns 6.25% interest compounded quarterly. She made no other deposits or withdrawals. What was the amount in her account at the end of 6 months. ______________

3. Ronda Perrin had $5050 deposited at City National Bank on June 1. The money earns interest at a rate of 5.25% compounded quarterly. What is the amount in the account on March 1 of the following year if no deposits or withdrawals were made? ______________

4. Delbert Renfer deposited $2400 in a new savings account on March 1. The savings account earns 6.0% interest compounded monthly. How much was in the account on June 1? ______________

5. Jeanne Crawford had $9675.95 deposited in an account paying $6\frac{1}{4}$% interest compounded semiannually. How much would Jeanne have in her account two years later? ______________

6. You deposit $2500 in a special savings account. The account earns interest at a rate of 6.25% compounded monthly. What amount will be in your account at the end of 5 months if no deposits or withdrawals are made? ______________

7. You deposit $500 in an account paying 7% compounded monthly. Two months later you deposit another $500 in the account. What amount will be in your account 6 months from the original investment? ______________

8. Your bank pays 8% compounded monthly. You deposit $100 per month. What amounts do you have in your account after 5 months if no withdrawals are made? ______________

Copyright © by Glencoe Division

Name ______________________ Date ______________

LESSONS 4-7, 4-8

Compound Interest Using Tables and Daily Compounding Using Tables

To compute compound interest quickly, you can use a compound interest table which shows the amount of $1.00 for many interest rates and interest periods. To use the table, you must know the total number of interest periods and the interest rate per period. Usually the more frequently interest is compounded, the more interest you will earn.

AMOUNT = ORIGINAL PRINCIPAL × AMOUNT OF $1.00

COMPOUND INTEREST = AMOUNT − ORIGINAL PRINCIPAL

Use the compound interest tables below to solve. Round answers to the nearest cent.

AMOUNT OF $1.00			
TOTAL INTEREST PERIODS	INTEREST RATE PER PERIOD		
	1.250%	1.375%	1.500%
1	1.01250	1.01375	1.01500
2	1.02516	1.02769	1.03023
3	1.03797	1.04182	1.04568
4	1.05095	1.05614	1.06136
5	1.06408	1.07067	1.07728
6	1.07738	1.08539	1.09344
7	1.09085	1.10031	1.10984
8	1.10449	1.11544	1.12649

AMOUNT OF $1.00 AT 5.5% COMPOUNDED DAILY, 365-DAY YEAR					
DAY	AMOUNT	DAY	AMOUNT	DAY	AMOUNT
21	1.00316	30	1.00452	80	1.01212
22	1.00331	40	1.00604	90	1.01364
23	1.00347	50	1.00755	100	1.01517
24	1.00362	60	1.00907	120	1.01823
25	1.00377	70	1.01059	140	1.02131

1. Valley Savings and Trust pays 5% interest compounded quarterly on regular savings accounts. John Hamilton deposited $2000 in a regular savings account for $1\frac{1}{2}$ years. He made no other deposits or withdrawals during the period. How much interest did John's money earn? ______________

2. Charles Johnson deposited $4400 in a savings account earning 6% compounded quarterly. If he makes no other deposits or withdrawals, how much will his money earn in 2 years? ______________

3. On January 4, Janelle Ruskinoff deposited $2192.06 in a savings account that pays 5.5% interest compounded daily. How much will Janelle's money earn in 24 days? ______________

4. Alfred Russ has a savings account at City Savings Bank. The account earns 5.5% interest compounded daily. On February 2, the amount in his account was $580. How much will be in the account in 40 days? ______________

5. On October 1, Joanne Booth deposited $1120 in a savings account that pays 5.5% interest compounded daily. On October 22, how much interest had been earned on the principal in Joanne's account? ______________

6. You have a savings account at Federal Savings. The account earns 5.5% interest compounded daily. On March 6, you had $1645.72 in your account. How much would be in the account on July 4? ______________

7. A trust fund of $5000 was started on January 1, 1980. The fund earned 5.5% interest compounded quarterly until January 1, 1982. Then the fund earned 15% interest compounded monthly until July 1, 1982. Then it earned 1.5% interest compounded annually until July 1, 1987. Finally, the fund will earn 5.5% interest compounded daily. What will the fund be worth on November 18, 1987? ______________

Copyright © by Glencoe Division

Name ______________________ Date ______________

LESSONS 5-1, 5-2

Sales Tax and Total Purchase Price

Many state, county, and city governments charge a sales tax on certain items or services that you buy. The sales tax rate is usually expressed as a percent. Most stores give you a sales receipt as proof of purchase. The receipt shows the selling price of each item or service you purchased, the total selling price, the sales tax (if any), and the total purchase price.

SALES TAX = SALES TAX RATE × TOTAL SELLING PRICE

TOTAL PURCHASE PRICE = TOTAL SELLING PRICE + SALES TAX

---

1. Erika Long purchases a $459.25 color TV set. The sales tax rate is 6%. What is the sales tax? ______

2. At R&J Department Store, Luke Russell purchases a stereo/cassette tape player for $164.99 and 3 tapes for $8.98 each. The sales tax rate is 6.0%. What is the sales tax on Luke's purchases? ______

3. Agnes Henderson purchased the following school supplies: 4 pens at 98¢ each, 5 packs of pencils at 88¢ each, 4 notebooks at $1.69 each, and 4 packs of filler paper at 79¢ each. She also purchased a calculator for $8.95. The sales tax rate is 7.5%. What is the total purchase price? ______

4. Ling Garn purchased 4 tires at $129.99 each, 2 sheepskin seatcovers at $59.89 each, a battery for $79.88, a muffler for $37.50, and 4 shock absorbers at $19.95 each. The sales tax rate is 5.75%. What is the total purchase price of Ling's purchases? ______

5. You purchase a $205.00 camera, a case for $23.89 a wide-angle lens for $127.88, and a zoom lens for $189.89. The sales tax rate is 6.0%.

a. What is the total selling price of your purchases? ______

b. What is the sales tax on your purchases? ______

c. What is the total purchase price of your purchases? ______

Complete the following sales receipts.

6.

Foil	2.88
Tissue	1.83
Aspirin	3.59
Cat food	6.51
Bleach	.97
Pantyhose	3.89
Towels	1.89
Soap	1.59
SUBTOT	______
TX 6%	______
TOT PUR PR	______

7.

Paint	18.90
Paint	18.90
Paint	18.90
Brush	7.89
Brush	3.29
Roller	9.49
Thinner	4.39
Dropcloth	2.29
SUBTOT	______
TX 5.5%	______
TOT PUR PR	______

8.

Blouse	39.98
Shirt	24.00
Skirt	54.98
Shoes	39.98
Slacks	49.95
Tie	16.50
Socks	6.00
Belt	17.00
SUBTOT	______
TX 8%	______
TOT PUR PR	______

Copyright © by Glencoe Division

Name ______________________ Date ____________

LESSONS 5-3, 5-4

Unit Pricing and Finding the Better Buy

Many grocery stores give unit price information for the products they sell. You can use this information to determine which size of a product is the better buy based solely on price. The unit price of an item is its cost per unit of measure or count.

$$\text{UNIT PRICE} = \frac{\text{PRICE PER ITEM}}{\text{MEASURE OR COUNT}}$$

1. Willis Rusch recently purchased a 48-oz can of shortening. The total purchase price was $2.99. What was the price per ounce? ____________

2. Leslie Thompson sees that tea bags are on sale for $2.49 for a 100-count box. What is the price per bag? ____________

3. Ron Wilcox wants to purchase some pickles. An 18-oz jar costs $2.49 while a 48-oz jar costs $6.60.

   a. What is the unit price of each? ____________

   b. Which is the better buy? ____________

4. Connie Stafford wants to purchase some pineapple. A 12-oz can costs $0.79 while an 18-oz can costs $1.04.

   a. What is the unit price of each? ____________

   b. Which is the better buy? ____________

5. The Food Town Store sells sandwich bags. The 75-count box costs $0.79. The 125-count box costs $1.29.

   a. What is the unit price of each? ____________

   b. Which is the better buy? ____________

6. The Shopperland Supermarket sells 3 sizes of the same brand of pork and beans. The 16-oz cans are priced 3 for $1.39; the 32-oz cans are priced 2 for $1.89; the 54-oz cans cost $1.61 each. Which is the best buy? ____________

7. You are shopping for the best buy on your favorite brand of instant mashed potatoes. The potatoes are available in an 8-oz bag for $0.89, a 16-oz bag for $1.47, or a 32-oz bag for $2.99. Which size is the best buy? ____________

8. Plastic wrap can be purchased for $1.09 for 100 feet, $1.55 for 200 feet, or $2.09 for 275 feet. Which is the best buy? ____________

9. You are purchasing cotton swabs that come in 2 different sized containers. One box has 300 swabs for $1.99 and the other box is specially marked 30% more swabs for the same price. How much will you save per swab by buying the larger box? ____________

Copyright © by Glencoe Division

Name ______________________ Date ______________

LESSON 5-5

Coupons and Rebates

Manufacturers, stores, and service businesses offer customers discounts through *coupons* and *refunds* or *rebates*. Manufacturer's or store coupons are redeemed at the time of purchase. Manufacturers rebates are obtained by mailing in a rebate coupon along with the sales slip and the U.P.C. (Universal Product Code) label from the product.

FINAL PRICE = TOTAL SELLING PRICE − TOTAL SAVINGS

1. Jese Aranud purchased a 32-oz bottle of liquid rubber cement for $8.19. He had the store coupon for $1.00. What is the final price of the rubber cement? ______

2. Bertha DeVys had the oil changed in her company car. She had a $3.00 off coupon from the Grease Monkey oil change service. The regular price of an oil change is $22.99. What is the final price? ______

3. Albert Poole purchased an AT&T phone answering system for his home office. The list price was $99.86. He also purchased a 12-foot handset cord for $8.95 and 50 feet of telephone wire for 15¢ a foot. He received a $5.00 rebate on the phone system and had a store coupon for $1.00 off each accessory. If he paid a 6% sales tax on the final price, how much did he pay? ______

4. A Thanksgiving Day promotion included a $1.00 mail-in rebate for the purchase of a turkey weighing at least 20 lbs. Esther Rothert purchased a 22-lb turkey for 89¢ a pound. What is the price after the rebate if an envelope costs 15¢ and a stamp costs 29¢. ______

5. A Fourth of July promotion included a $5.00 mail-in rebate for the purchase of a picnic cooler and a store coupon for 50¢ off the price of a case of 24 cans of soft drinks. For the company picnic Carl Rhiel purchased a 48-quart cooler for $32.99 and a case of pop for $6.99. What is the price after the rebate if an envelope costs 15¢ and a stamp costs 29¢? ______

6. A box of ten $3\frac{1}{2}$-inch high density computer disks costs $15.90 and has a $2.00 mail-in rebate on the back of the box. What is the price per disk after the rebate if an envelope costs 20¢ and a stamp costs 29¢? ______

7. If you purchase a toner cartridge for a laser printer and a case of 10 reams of laser paper, you get a mail-in coupon and can receive $1.50 back by mail. If an envelope costs 18¢ and a stamp costs 29¢, how much is your actual rebate? ______

Copyright © by Glencoe Division

Name ______________________________ Date ______________

LESSONS 5-6, 5-7

Markdown and Sale Price

Stores often sell products at sale prices that are lower than their regular selling prices. The markdown or discount is the amount of money that you save by purchasing a product at the sale price. The markdown rate or discount rate of an item is its markdown expressed as a percent of its regular selling price.

MARKDOWN = REGULAR SELLING PRICE − SALE PRICE

MARKDOWN = MARKDOWN RATE × REGULAR SELLING PRICE

SALE PRICE = REGULAR SELLING PRICE − MARKDOWN

For problems 1 and 2, determine the markdown.

1. **2C. Button front tank.** Ribbed cotton in blue, olive or yellow. S-M-L. Orig**. $29, **sale $20.30** ______________

2. **2D. Tattersall blouse.** Roll sleeves, patch pockets. Cotton; sizes 4-14. Reg. $48, **sale $36** ______________

For problems 3 and 4, determine the markdown and the sale price.

3. Petites Coordinates
25% Off Reg. 40.00
Blazers, skirts & pants in navy, black or tan. Sizes 6–16.

______________ ______________

4. Spring Jackets
up to 30% Off
Wide range of basic and fashion colors in nylon or poplin jackets. S, M, L, XL. Reg. 59.99

______________ ______________

5. Wicker Book Store has paperback books marked down 25%. What is the total sale price of 4 books that regularly sell for $8.95, $9.55, $6.85, and $12.95? ______________

6. During a September Sale, Kauffman Kamper Supply has marked down all items 30%. Judy Vaughn purchases a deluxe portable grill that regularly sells for $79.99 and a set of barbecue tools that regularly sells for $19.98. What is the total sale price? ______________

7. Lillian Morrin shopped at Bauman's Department Store during the end-of-summer sale. All luggage was marked down 35%. She purchased a suitcase that regularly sells for $74.95, a garment bag that regularly sells for $80.00, and a tote bag that regularly sells for $39.99. What is the total sale price Lillian paid? ______________

8. During the Winter Clearance Sale, Townsend Cyclery has marked down all bicycles 20% and all accessories 25%. You purchase a Super Sport 26-inch 10-speed bike that regularly sells for $199.95, a chain and lock that regularly sell for $16.99, a tire gauge and tire pump that regularly sell for $24.49, and a deluxe generator and light package that regularly sells for $29.95. What is the total sale price for your purchases? ______________

Copyright © by Glencoe Division

Name ______________________ Date ______________

LESSON 6-1

Sales Receipts

When you make a purchase with a credit or charge card, the salesclerk prepares a sales receipt. The receipt shows your name and account number, the price of each item you purchased, the sales tax, and the total purchase price.

TOTAL PURCHASE PRICE = TOTAL SELLING PRICE + SALES TAX

Complete the sales receipt.

1.

DATE	AUTH. NO.	INDENTIFICATION	CLERK	REG/DEPT	☑ TAKE ☐ SEND
	15		3L		

QTY.	CLASS	DESCRIPTION	PRICE	AMOUNT	
3		pr. socks	2.49	7	47
2		shoes	39.95	79	90
1		Heelguard	3.49	3	49
The issuer of the card identified on this item is authorized to pay the amount shown as TOTAL upon proper presentation. I promise to pay such TOTAL (together with any other charges due thereon) subject to and in accordance with the agreement governing the use of such card.			SUB TOTAL	90	86
CUSTOMER SIGNATURE X			TAX	5	45
SALES SLIP			TOTAL		

2.

12 4718 17 Alice May R & W SERVICE		DATE 2/20/–			
		Quantity	Price	Amount	
Products & Services	SUPREME ☐ REGULAR ☐ UNLEADED ☑	20.4	1.30		
	Motor Oil	1 qt.	1.45	1	45
	Antifreeze	1 g		4	95
	Spark Plugs	6	1.09		
SIGN HERE X		Sales Tax		2	37
		Total			

3. Ian Bollinger purchases a picnic table for $95.50, a chaise lounge for $21.90, and a lawn chair for $14.98. There is a sales tax of 6.5%. He charges the purchases to his bank charge card. What is the total purchase price? ______________

4. Robin Hankinson uses her bank charge card to purchase a storage shed. The shed costs $495.95 plus 5.75% sales tax. What is the total purchase price on Robin's sales receipt? ______________

5. Complete the sales receipt for 5 rosebushes at $6.95 each, a box of rose food for $7.75 and a lawn rake for $12.95. The sales tax rate is 5.75%.

DATE	AUTH. NO.	INDENTIFICATION	CLERK	REG/DEPT	☐ TAKE ☐ SEND
10/21			UM		

QTY.	CLASS	DESCRIPTION	PRICE	AMOUNT	
The issuer of the card identified on this item is authorized to pay the amount shown as TOTAL upon proper presentation. I promise to pay such TOTAL (together with any other charges due thereon) subject to and in accordance with the agreement governing the use of such card.			SUB TOTAL		
CUSTOMER SIGNATURE X			TAX		
SALES SLIP			TOTAL		

6. Complete the sales receipt for 2 sweaters at $29.99 each, 3 ties at $16.99 each, 5 pairs of socks at $3.65 each, and a belt for $16.79. The sales tax rate is 7.25%.

DATE	AUTH. NO.	INDENTIFICATION	CLERK	REG/DEPT	☐ TAKE ☐ SEND
10/21			UM		

QTY.	CLASS	DESCRIPTION	PRICE	AMOUNT	
The issuer of the card identified on this item is authorized to pay the amount shown as TOTAL upon proper presentation. I promise to pay such TOTAL (together with any other charges due thereon) subject to and in accordance with the agreement governing the use of such card.			SUB TOTAL		
CUSTOMER SIGNATURE X			TAX		
SALES SLIP			TOTAL		

Copyright © by Glencoe Division

Name ______________________ Date ______________

LESSON 6-2

Account Statements

When you have a credit card or charge account, you receive a monthly statement. The statement lists all transactions that were processed by the closing date for that month. If your previous bill was not paid in full, a finance charge is added. The finance charge is interest that is charged for delaying payment.

NEW BALANCE = PREVIOUS BALANCE + FINANCE CHARGE + NEW PURCHASES − (PAYMENTS + CREDITS)

1. What is the new balance for the credit statement shown?

BILLING DATE	PREVIOUS BALANCE	FINANCE CHARGE	PAYMENTS & CREDITS	NEW PURCHASES	NEW BALANCE
10/1/–	$139.50	$2.32	$45.00	$29.98	

2. What is the new balance for the credit statement shown?

BILLING DATE	PREVIOUS BALANCE	FINANCE CHARGE	PAYMENTS & CREDITS	NEW PURCHASES	NEW BALANCE
12/11/–	$185.74	$2.71	$70.94	$49.80	

3. You received this monthly statement from Bank Card. What is your new balance?

BILLING DATE	PREVIOUS BALANCE	FINANCE CHARGE	PAYMENTS & CREDITS	NEW PURCHASES	NEW BALANCE
09/22/–	$374.06	$6.55	$10.00	$144.99	

4. Complete the account statement. Previous balance of $716.45; payments of $150 and $75; new purchases of $29.98, $129.90, and $10.46; finance charge of $12.54.

BILLING DATE	PREVIOUS BALANCE	FINANCE CHARGE	PAYMENTS & CREDITS	NEW PURCHASES	NEW BALANCE
02/1/–					

5. Complete the account statement. Previous balance of $78.80; payment of $78.80; new purchases of $24.60 and $54.98, no finance charge.

BILLING DATE	PREVIOUS BALANCE	FINANCE CHARGE	PAYMENTS & CREDITS	NEW PURCHASES	NEW BALANCE
04/1/–					

6. Complete the account statement. Previous balance of $410.91; payments of $150 and $150; return credit of $21.90; new purchases of $71.80, $21.90, $116.60, $10.49, $51.80, and $6.75; finance charge of $7.19.

BILLING DATE	PREVIOUS BALANCE	FINANCE CHARGE	PAYMENTS & CREDITS	NEW PURCHASES	NEW BALANCE
06/1/–					

Copyright © by Glencoe Division

Name ______________________ Date ______________

LESSONS 6-3, 6-4

Finance Charge — Previous-Balance and Unpaid-Balance Method

Some credit card companies use the previous-balance method to compute finance charges. They compute the finance charge based on the amount you owed on the closing date of your last statement. The periodic rate is the monthly finance charge rate. Other companies use the unpaid-balance method. They compute the finance charge based on that portion of the previous balance that you have not paid.

UNPAID BALANCE = PREVOUS BALANCE − (PAYMENTS + CREDITS)

FINANCE CHARGE = PERIODIC RATE × UNPAID BALANCE

NEW BALANCE = PREVIOUS BALANCE + FINANCE CHARGE + NEW PURCHASES − (PAYMENTS + CREDITS)

---

1. Rosie Lane has a charge account at the Cosmopolitan Department Store where the periodic rate is 1.58%. A portion of her account statement is shown. Find the finance charge and new balance using:

a. previous-balance method **b.** unpaid-balance method

	BILLING DATE	PREVIOUS BALANCE	FINANCE CHARGE	PAYMENTS & CREDITS	NEW PURCHASES	NEW BALANCE
a.	05/20/–	$194.06		$61.50	$29.80	
b.	05/20/–	$194.06		$61.50	$29.80	

2. You have a charge account with a periodic rate of 2.08%. Your monthly statement shows purchases totaling $416.49 and a payment of $750. Complete this portion of your account statement and find the new balance using:

a. previous-balance method **b.** unpaid-balance method

	BILLING DATE	PREVIOUS BALANCE	FINANCE CHARGE	PAYMENTS & CREDITS	NEW PURCHASES	NEW BALANCE
a.	02/01/–	$981.35				
b.	02/01/–	$981.35				

3. The periodic rate is 2.0%; previous balance is $291.84; payment of $200; new purchases of $17.40, $17.70, and $46.04. Complete the account statements using:

a. previous-balance method **b.** unpaid-balance method

	BILLING DATE	PREVIOUS BALANCE	FINANCE CHARGE	PAYMENTS & CREDITS	NEW PURCHASES	NEW BALANCE
a.	07/01/–					
b.	07/01/–					

Copyright © by Glencoe Division

Name ________________________________ Date ______________

LESSONS 6-5, 6-6

Finance Charge — Average Daily Balance

Many companies calculate the finance charge using the average-daily-balance method where no new purchases are included. The average daily balance is the average of the account balance at the end of each day of the billing period. New purchases posted during the billing period may or may not be included when figuring the balance at the end of the day. The finance charge is calculated by multiplying the periodic rate by the average daily balance.

$$\text{AVERAGE DAILY BALANCE} = \frac{\text{SUM OF DAILY BALANCES}}{\text{NUMBER OF DAYS}}$$

FINANCE CHARGE = PERIODIC RATE × AVERAGE DAILY BALANCE

NEW BALANCE = UNPAID BALANCE + FINANCE CHARGE + NEW PURCHASES

1.

Billing Periods	Payment	End-of-day Balance	Number of Days	Sum of Balances
9/01-9/10		$410.20	10	$4102.00
9/11	$150.00	260.20	1	260.20
9/12-9/30		260.20	19	
		TOTALS		

What is the average daily balance without new purchases? ____________

2. Compute the average balance, finance charge, and new balance as of July 1 using the average-daily-balance method where no new purchases are included. The periodic rate is 1.4%.

Date	Transaction	Amount
June 1	Balance	$800.00
June 11	Payment	200.00
June 25	Purchase	150.00

3. A portion of your account statement for November from Charge-All Credit Company is shown. The finance charge is computed using the average-daily-balance method where new purchases are included. Find the average daily balance, the finance charge, and the new balance.

REFERENCE	POSTING DATE	TRANSACTION DATE	DESCRIPTION	PURCHASES & ADVANCES	PAYMENTS & CREDITS
001781	3/12		PAYMENT		45.00
007619	3/21	3/17	CornerDrug Co.	31.75	

BILLING PERIOD	PREVIOUS BALANCE	PERIODIC RATE	AVERAGE DAILY BALANCE	FINANCE CHARGE
3/05-4/04	$161.43	2.0%		

PAYMENTS & CREDITS	PURCHASES & ADVANCES	NEW BALANCE	MINIMUM PAYMENT	PAYMENT DUE
			$20.00	3/24

Copyright © by Glencoe Division

Name ______________________ Date ______________

Charge Accounts: A Simulation

You receive your monthly statement from the bank that issued your bank credit card. A portion of the credit card statement is shown.

Reference	Posting Date	Transaction Date	Description	Purchases & Advances	Payments & Credits
142116	7/7	7/1	Village Sports	$98.20	
132157	7/17	7/12	Adams Ins.	$292.80	
270-32	7/22		PAYMENT		$100.00
122384	7/28	7/24	Ace Tire Co.	$84.80	

Billing Period	Prev. Balance	Periodic Rate	FINANCE CHARGE
7/1-7/31	$186.80	1.5%	

1. If your credit card company used the previous-balance method of computing the finance charge, they would compute the charge on the amount you owed on the closing date of your last statement. Complete the statement below to determine the finance charge and the new balance.

Billing Period	Prev. Balance	Periodic Rate	FINANCE CHARGE
7/1-7/30	$186.80	1.5%	

Payments & Credits	Purchases & Advances	New Balance	Minimum Payment	Payment Due
			$20.00	8/25

2. If your credit card company used the unpaid-balance method of computing the finance charge, they would compute the charge on the portion of the previous balance that you have not paid. Complete the statement below to determine the unpaid balance, the finance charge, and the new balance.

Billing Period 7/1–7/30

Previous Balance	Payments & Credits	Unpaid Balance	Finance Charge	New Purchases	New Balance
$186.80	$100.00				

Copyright © by Glencoe Division

Name ____________________ Date ____________

Charge Accounts: A Simulation
(CONTINUED)

3. If your credit card company used the average-daily-balance—no new purchases included method of computing the finance charge, the charge on your average daily balance would be computed without including new purchases. Complete the statement below to determine the average daily balance—no new purchases included, the finance charge, and the new balance.

BILLING PERIOD: 7/1 TO 7/30					
PREVIOUS BALANCE	PAYMENTS & CREDITS	AVG DAILY BALANCE NO NEW PURCHASES	FINANCE CHARGE	NEW PURCHASES	NEW BALANCE

4. Assume that your credit card company computed the finance charge using the average-daily-balance method where new purchases *are* included.

a. Use the chart below to find the average daily balance.

Dates	Purchase or Payment	End-of-Day Balance	Number of Days	Sum of Balances
7/1-6				
7/7	$ 98.20			
7/18-16				
7/17	$292.80			
7/18-21				
7/22	$100.00			
7/23-27				
7/28	$ 84.80			
7/29-31				
Total				
Average Daily Balance				

b. Complete the statement below to show the average daily balance, the finance charge, and the new balance.

Billing Period	Prev. Balance	Periodic Rate	Average Daily Balance	Finance Charge
7/1–7/31	$186.80	1.5%		
Payments & Credits	**Purchases & Advances**	**New Balance**	**Minimum Payment**	**Payment Due**
			$40.00	8/25

5. If you had your choice, which method would you prefer that the bank use to figure your finance charge? ____________

Copyright © by Glencoe Division

Name ______________________ Date __________

LESSON 7-1

Single-Payment Loans

A single-payment loan is a loan that you repay with one payment after a specified period of time or term. Ordinary interest is calculated by basing the term on a 360-day year. Exact interest is calculated by basing the term on a 365-day year. The maturity value of the loan is the total amount you repay.

MATURITY VALUE = PRINCIPAL + INTEREST OWED

---

1. Tao Bergolt's bank granted him a single-payment loan of $4400 at an interest rate of 12%. The term of the loan is 172 days. What is the maturity value of his loan at exact interest? __________

2. Jane Dimas obtained a single-payment loan of $420 to pay a repair bill. She agreed to repay the loan in 90 days at an interest rate of 12.75%, ordinary interest. What is the maturity value of her loan? __________

3. Joyce Stein borrowed $8460 from Merchants Trust Company to pay for some merchandise for her dress shop. The loan is for 45 days at 12.75% exact interest. What is the maturity value of the loan? __________

4. Gardening, Inc., borrowed $94,500 at 11.65% ordinary interest for 15 days. What is the maturity value? __________

5. Ruth and Juan Dimas would like to borrow $2600 for 90 days to pay their real estate tax. State Savings and Loan charges 14.00% ordinary interest while Security Bank charges 14.25% exact interest.

State __________

a. What is the maturity value of each loan?

Security __________

b. Where should they borrow the money? __________

6. The Walker Trust Company charges exact interest, while Farmers and Merchants Savings Bank charges ordinary interest. You plan to borrow $9000 for 60 days at 14%. What is the cost of interest at each bank? Which bank offers you the better deal?

Walker __________

F&M __________

____

7. You have a chance to loan $6500 at 12.65% interest for 95 days.

a. If you charge ordinary interest, how much will you earn? __________

b. If you charge exact interest, how much will you earn? __________

c. Suppose you were borrowing the money, would you prefer to pay ordinary interest or exact interest? Why? __________

Copyright © by Glencoe Division

Name ______________________________ Date ______________

LESSONS 7-2, 7-3

Installment Loans—Amount Financed, Finance Charge

An installment loan is repaid in several equal payments over a specified period of time. Usually, you make a down payment to cover a portion of the cash price of the item. The amount you finance is the portion of the cash price that you owe after making the down payment.

AMOUNT FINANCED = CASH PRICE − DOWN PAYMENT

MONTHLY PAYMENT = (AMOUNT OF LOAN ÷ 100) × MONTHLY PAYMENT FOR $100 LOAN

TOTAL AMOUNT REPAID = NUMBER OF PAYMENTS × MONTHLY PAYMENT

FINANCE CHARGE = TOTAL AMOUNT REPAID − AMOUNT FINANCED

---

1. Alan Lewis purchased a new computer for his office using the store's installment credit plan. The computer cost $5991.64. What amount did Adam finance if he made a 40% down payment? ______________

2. Lloyd and Linda Pearl want to remodel the dining room in their house. They would like to pay 30% of the cost first and repay the amount financed in installments. The estimated cost for this job is $6890. They can borrow the money at 15% for 48 months. What is

a. the down payment? ______________

b. the amount financed? ______________

c. the payment? ______________

d. the finance charge? ______________

3. The Potters obtained an installment loan of $5000 from the credit union to pay for their son's tuition. They obtained the loan at an APR of 10% and agreed to repay the loan in 12 months. What is the finance charge? ______________

MONTHLY PAYMENT ON A $100 LOAN				
Term in Months	Annual Percentage Rate			
	10%	12%	15%	18%
6	$17.16	$17.25	$17.40	$17.55
12	8.79	8.88	9.03	9.17
18	6.01	6.10	6.24	6.38
24	4.61	4.71	4.85	4.99
30	3.78	3.87	4.02	4.16
36	3.23	3.32	3.47	3.62
42	2.83	2.93	3.07	3.23
48	2.54	2.63	2.78	2.94

4. Allison O'Conell would like an installment loan for $500. Her bank will loan her the money at 18% for 18 months. Her insurance company will loan her the money at 15% for 24 months. Which loan would cost her less? ______________

5. You want to borrow $1500 to take a vacation. First Century Bank will lend you the money at 15% for 12 months. Fidelity Savings and Loan will lend you the money at 10% for 24 months.

a. How much will each loan cost? ______________

b. Which loan would cost you less? ______________

c. How much less would it cost? ______________

Copyright © by Glencoe Division

Name ______________________ Date ______________

Installment Loan—Monthly Payment Allocation—Paying Off

Part of each monthly payment is used to pay the interest on the unpaid balance of the loan and the remaining part is used to reduce the balance. To pay off a loan before the end of the term you must pay the previous balance plus the current month's interest.

PAYMENT TO PRINCIPAL = MONTHLY PAYMENT − INTEREST
NEW PRINCIPAL = PREVIOUS PRINCIPAL − PAYMENT TO PRINCIPAL
FINAL PAYMENT = PREVIOUS BALANCE + CURRENT MONTH'S INTEREST

---

1. Ralph Phillips obtained a 12-month, $1500 loan at 12% from his credit union. His monthly payment is $133.20. For the first payment:

a. What is the interest? ______

b. What is the payment to principal? ______

c. What is the new balance? ______

2. Rita Rodriguez obtained a 24-month, $8500 loan at 8% from Tri-County Savings & Loan. Her monthly payment is $384.20. For the first payment:

a. What is the interest? ______

b. What is the payment to principal? ______

c. What is the new balance? ______

3. Complete the repayment schedule for a $2400 loan at 12% for six months.

Payment Number	Payment	Amount For Interest	Amount For Principal	New Principal
1	$414.00	$24.00	$390.00	$2010.00
2	$414.00	$20.10		
3	$414.00			
4	$414.00			
5	$414.00			
6				

4. Patricia Nichols took out a $4000 simple interest loan at 12% for 12 months. After 5 payments the balance was $2392.16. She pays off the loan when the next payment is due.

a. What is the current month's interest? ______

b. What is the final payment? ______

5. Chad Roth took out a $9100 simple interest loan at 10% for 36 months. After 27 payments the balance is $2526.85. He pays off the loan when the next payment is due.

a. What is the current month's interest? ______

b. What is the final payment? ______

Copyright © by Glencoe Division

Name ______________________ Date ______________

LESSON 7-6

Determining the APR

If you know the number of monthly payments you will make and the finance charge per $100 of the amount financed, you can find the annual percentage rate (APR) of the loan from a table. You can use the APR to compare the relative cost of borrowing money.

$$\text{FINANCE CHARGE PER \$100} = \$100 \times \frac{\text{FINANCE CHARGE}}{\text{AMOUNT FINANCED}}$$

Use the APR table below to solve.

	ANNUAL PERCENTAGE RATES										
APR	**10.00%**	**10.25%**	**10.50%**	**10.75%**	**11.00%**	**11.25%**	**11.50%**	**11.75%**	**12.00%**	**12.25%**	**12.50%**
Term	**Finance Charge Per $100 of Amount Financed**										
6	$ 2.94	$ 3.01	$ 3.08	$ 3.16	$ 3.23	$ 3.31	$ 3.38	$ 3.45	$ 3.53	$ 3.60	$ 3.68
12	5.50	5.64	5.78	5.92	6.06	6.20	6.34	6.48	6.62	6.76	6.90
18	8.10	8.31	8.52	8.73	8.93	9.14	9.35	9.56	9.77	9.98	10.19
24	10.75	11.02	11.30	11.58	11.86	12.14	12.42	12.70	12.98	13.26	13.54

1. Ruby Crye obtained an installment loan of $3870 to purchase new furniture. The finance charge is $378.10. She agreed to repay the loan in 18 monthly payments. What is the annual percentage rate? ________

2. Gordon Stewart obtained an installment loan of $10,900 to pay his daughter's tuition. The finance charge is $1172. He agreed to repay the loan in 24 monthly payments. What is the APR? ________

3. The Gomez family has just purchased a $2574.54 microcomputer. They made a down payment of $574.54. Through the store's installment plan, they have agreed to pay $121 a month for the next 18 months.

a. What is the amount financed? ________

b. What is the finance charge? ________

c. What is the APR? ________

4. Ellen Andrzejewski purchased the car in the ad. She made a down payment of $775. She financed the remainder at $331.16 a month for 24 months. What is the APR? ________

> **Jetta 4 Dr**
> #6840–5 speed, alloy wheels,
> cloth seats, stereo prep,
> Was $8925 NOW $7775

5. You are buying a color television set that costs $964.91. To use the installment plan available at the department store, you must make a down payment of 20% and make 12 monthly payments of $68.68 each.

a. What is the amount financed? ________

b. What is the finance charge? ________

c. What is the APR for your loan? ________

Copyright © by Glencoe Division

Name ______________________________ Date ______________

LESSON 7-7

Refund of Finance Charge

If you repay an installment loan before the final due date, you may be entitled to a refund of part of the finance charge. Your refund is usually stated as a percent of the finance charge. The percent of refund may be determined by using a table, called the rebate schedule.

REFUND = PERCENT REFUND × FINANCE CHARGE

Use the table of percent refund below to solve.

TERM OF LOAN	NUMBER OF MONTHS LOAN HAS RUN													
	1	2	3	4	5	6	7	8	9	10	11	12	13	14
3	50.00	16.67	0											
6	71.43	47.62	28.57	14.29	4.76	0								
9	80.00	62.22	46.67	33.33	22.22	13.33	6.67	2.22	0					
12	84.62	70.51	57.69	46.15	35.90	26.92	19.23	12.82	7.69	3.85	1.28	0		
15	87.50	75.83	65.00	55.00	45.83	37.50	30.00	23.33	17.50	12.50	8.33	5.00	2.50	0.83
18	89.47	79.53	70.18	61.40	53.22	45.61	38.60	32.16	26.32	21.05	16.37	12.28	8.77	5.85

1. Bernard Hammond had a 15-month installment loan, which he repaid in 12 months. The total finance charge was $180. How much was Bernard's refund? ______________

2. Betty Long took a 12-month installment loan to help pay for her television set. The total finance charge was $72.25. Betty repaid the loan in 4 months. How much was Betty's refund? ______________

3. Martha Conway obtained an installment loan of $1450 from her bank. She agreed to pay $128.79 a month for the next 12 months. The finance charge totaled $95.48. How much of the finance charge can Martha save by repaying the loan in 8 months? ______________

4. Steve Krzyzewski obtained a loan of $9400 on a used car. He agreed to pay $556.17 a month for 18 months. How much is the finance charge? How much is the refund if he repays the loan in 12 months? ______________

5. You have decided to put a new roof on your house. You obtained an installment loan of $3500 through your bank and agreed to repay the loan in 18 monthly payments of $213.44.

a. What is the finance charge? ______________

b. How much can you save by repaying the loan in 9 months instead of making all 18 payments? ______________

6. Bob and Beth Green want to repaint the exterior of their home. They obtained an installment loan of $2400 through the bank and agreed to repay the loan in 18 monthly payments of $144.57 each. What is the finance charge on the loan? ______________ How much can they save by repaying the loan in 6 months instead of making all 18 payments? ______________

Copyright © by Glencoe Division

Name ______________________________ Date ______________

LESSONS 8-1, 8-2

Purchasing a New Automobile, Sticker Price, and Dealer's Cost

An automobile's sticker price shows all charges for the car, including the base price, all options, and the destination charge. Automobile dealers pay less than the prices on the sticker for both the basic automobile and the options.

STICKER PRICE = BASE PRICE + OPTIONS + DESTINATION CHARGE

DEALER'S COST = PERCENT OF BASE PRICE + PERCENT OF OPTIONS PRICE + DESTINATION CHARGE

1. Loren Weber wants to buy the Pantheon XE90 sports car listed below. There is a 10% sales tax on the purchase price of the car in Loren's state. If Loren pays the full sticker price for the car, what is the total price? ____________

CODE	DESCRIPTION	LIST PRICE
B19	PANTHEON XE90 SPORTS CAR	$8264.00
C21	METALLIC TRI/STRIPE	65.95
IL1	LEATHER UPHOLSTERY	621.40
AC3	AIR CONDITIONING	647.80
TG2	TINTED GLASS	79.60
S14	STEREO AM/FM CASSETTE	516.70
T14W	TIRES-WHITE SIDEWALL RADIAL	152.90
E41	DESTINATION CHARGE	246.83

2. Hillary Ellios wants to buy a Baxter XL-10 sports car. There is a 9% sales tax on the purchase price of the car in her state. If Hillary pays the full sticker price for the car, how much is the total? ____________

C/C	DESCRIPTION	LIST PRICE
A78	BAXTER XL-10 SPORTS CAR	$9215.00
C41	METALLIC BROWN	0.00
E30	LEATHER UPHOLSTERY	583.00
B45	AIR CONDITIONING	508.43
P26	TINTED GLASS	79.25
L04	STEREO, AM/FM RADIO	489.40
X-93	TIRES-WHITE SIDEWALL	63.80
M72	DESTINATION CHARGE	158.64

3. The Baker's are buying a new Galaxy I-90 sports compact that has a base price of $7995 with options totaling $2410.80 and a destination charge of $221.80. The dealer's cost is estimated to be 85% of the base price and 88% of the price of the options.

a. What is the sticker price of the car? ____________

b. What is the estimated dealer's cost? ____________

4. Rochell Holland is importing a new Fugi model car from Japan for her showroom. The base price or the car is $8425. The options total $1242, and the destination charge is $2695. The import duty is 35% of the dealer's cost. How much does Ruth pay if she is able to negotiate a 22% discount on both the base price and the options price? ____________

Copyright © by Glencoe Division

Name ________________________________ Date ____________

LESSON 8-3

Purchasing a Used Automobile

Automobile dealers usually advertise used cars for prices that are higher than what they expect you to pay. Used-car guides, published monthly, give the average prices for cars that were purchased from dealers during the previous month. This information can help you decide how much to offer for a used automobile.

AVERAGE RETAIL PRICE = AVERAGE RETAIL VALUE + ADDITIONAL OPTIONS − OPTIONS DEDUCTIONS − MILEAGE DEDUCTION

1. George Garcia owns a three-year-old Dugger that he wants to sell in order to buy a new Wabash. One used-car guide shows the average retail value of his car is $4100. He adds $50 for having a vinyl top, $100 for air conditioning, and $35 for power windows. He deducts $200 for excessive mileage. What is the average retail price he can ask as a selling price for his car? ____________

2. Yolanda Lovelace owns an Eaton II. She wants to sell it in order to buy the new Eaton III model. One used-car guide shows the average retail value of her car is $3500. She adds $50 for cruise control $200 for stereo AM/FM cassette/radio, and $175 for leather seat covers. She deducts $100 for no air conditioning and $100 for excessive mileage. What is the average retail price Yolanda can ask for her Eaton II? ____________

3. The Tookesons want to trade in their station wagon on the purchase of a new station wagon. A used-car guide shows the average trade-in value is $3610. They add $150 for air conditioning, $75 for overdrive, and $150 for the stereo cassette player. They subtract $100 for a manual transmission and $300 for excessive mileage. What is the average trade-in value of their car? ____________

4. Betty Flanagan is going to trade in her five-year-old Mystic Crown for a new automobile. A used-car guide shows the average trade-in value is $1275, plus $100 for air conditioning and $50 for an automatic transmission. She deducts $175 for excessive mileage. What is the average trade-in value of her car? ____________

Use the used-car guide at the right for problems 5 and 6.

5. You decide to purchase a used Mystic V8-PS-PE 11 Spt Cpe 2D with an AM/FM stereo/Tp, rear window defroster, and a tile steering wheel. The car does not have air conditioning. What is the average loan value of the car? ____________

6. You are going to trade in a Mystic PS-PE 12 Berlinetta Cpe. Your car has a T-Top, power door locks, power windows, power seats, and cruise control. It has a 4-cylinder engine with manual transmission. What is the average trade-in value of your car? ____________

Av'g Trd-in	BODY TYPE	Av'g Loan	Av'g Ret'l
	MYSTIC PS-PE		
8550	11 Spt Cpe 2D	7800	9670
9450	12 Berlinetta Cpe	8620	10575
10350	Spt Cpe Z28	9325	12650
500	**Add** T-Top	450	550
125	**Add** AM/FM Stereo	125	150
150	**Add** AM/FM Stereo.Tp	150	250
75	**Add** Power Door Locks	75	100
100	**Add** Power Windows	100	125
100	**Add** Power Seats	100	150
75	**Add** Rear Wind Defroster	75	150
75	**Add** Cruise Control	75	100
75	**Add** Tilt Strg Wheel	75	100
200	**Deduct** 4 Cyl. Engine	200	200
375	**Deduct** Manual Trans.	350	375
550	**Deduct** W/out Air Cond	500	525

Copyright © by Glencoe Division

Name ______________________________ Date ______________

LESSON 8-4

Automobile Insurance

Automobile insurance protects you against financial losses if your car is involved in an accident. Your annual premium depends on your base premium and your driver-rating factor. Your driver-rating factor depends on your age, your marital status, your sex, and the purpose for which you drive.

ANNUAL PREMIUM = ANNUAL BASE PREMIUM × DRIVER-RATING FACTOR

LIABILITY PREMIUM FOR A PRIVATE PASSENGER AUTOMOBILE

PROPERTY DAMAGE LIMITS	BODILY INJURY LIMITS					
	25/50	25/100	50/100	100/200	100/300	300/300
$ 25,000	$206.40	$218.80	$213.20	$252.00	$258.00	$286.80
50,000	212.40	224.80	237.20	258.00	264.00	293.20
100,000	220.80	233.20	245.60	266.40	272.40	301.20

PHYSICAL DAMAGE PREMUIM

COVERAGE	AGE GROUP	INSURANCE-RATING GROUP					
		10	11	12	13	14	15
Comprehensive $50-Deductible	A	$ 76.80	$ 81.60	$ 95.20	$108.00	$122.00	$135.60
	B	65.20	77.60	90.40	102.40	115.60	128.40
	C	62.00	74.00	86.00	98.00	110.40	122.80
	D	59.20	70.40	82.00	93.20	105.20	116.80
Collision $50-Deductible	A	$225.60	$246.00	$266.80	$287.20	$307.60	$328.00
	B	214.00	233.20	253.20	272.40	291.60	311.20
	C	204.00	222.80	241.60	260.00	278.40	296.80
	D	194.40	212.00	230.00	247.60	265.20	282.80

1. Julie Spiros, age 21 and unmarried, drives her car to and from work. Her driver-rating factor is 1.85. Her car is classified A-11. She has $25,000 property damage, 50/100 bodily injury, $50 deductible comprehensive, and $50 deductible collision coverages. What is her base premium? annual premium? ______ ______

2. Theo Norton is 18 years of age. His driver-rating factor is 2.30. The family car is a C-15. The insurance the Moores have is $50,000 property damage, 100/300 bodily injury, $50 deductible comprehensive, and $50 deductible collision. What is Tom's annual premium? ______

3. Amy Schorling, age 24, is unmarried and uses her car for business. Her driver-rating factor is 1.85. Her car is classified D-14. Her insurance consists of $25,000 property damage, 25/50 bodily injury, $50 deductible comprehensive, and no collision coverage.

a. What is Amy's annual base premium? ______

b. How much is her annual premium if she includes $50 deductible collision coverage? ______

4. Al Switlick, age 19 and unmarried, drives his C-11 car to and from work. His driver-rating factor is 4.10. His insurance includes $25,000 property damage, 25/100 bodily injury, $50 deductible comprehensive, and $50 deductible collision.

a. What is his annual premium? ______

b. What is his annual premium if he has the highest possible property damage and bodily injury coverage? ______

Copyright © by Glencoe Division

Name ______________________________ Date ______________

LESSON 8-5

Operating and Maintaining an Automobile

Many costs are involved in operating and maintaining an automobile. Variable costs increase the more you drive, while fixed costs remain about the same regardless of how many miles you drive.

$$\text{COST PER MILE} = \frac{\text{ANNUAL VARIABLE COST} + \text{ANNUAL FIXED COST}}{\text{NUMBER OF MILES DRIVEN}}$$

---

1. Tom Larson had fixed costs totaling $2805.60 last year. His variable costs totaled $1870.40. Arthur drove his auto 16,700 miles last year. What was his cost per mile? ______________

2. Joyce Staley drove 9910 miles last year. Her fixed costs totaled $1754.07 last year while her variable costs totaled $584.69. What was her cost per mile to the nearest tenth of a cent? ______________

3. John and Evelyn each own a car of the same model. John drove a total of 27,612 miles, while Evelyn drove 9821 miles. Both had fixed costs totaling $2357.04. John's variable cost totaled $4822.08, and Evelyn's totaled $589.26

a. How much money did John spend to operate and maintain his car on a cost-per-mile basis? ______________

b. How much did Evelyn spend per mile? ______________

4. Jill North drove her car about 7200 miles last year. Her fixed costs totaled $1058.40 and her variable costs were $1965.60. How much did it cost per mile for Jill to operate her car? ______________

5. Jim Spring maintained this record of his car expenses for last year: gas, $845.96; oil, lube, miscellaneous, $68.85; insurance, $418.50; and license, $35. He drove 8280 miles last year.

a. What were his total variable costs? ______________

b. What were his total fixed costs? ______________

c. What were his total costs? ______________

d. What was his cost per mile to the nearest tenth of a cent? ______________

6. Last year Wayne Johnson drove 21,986 miles and had these expenses: gas, $1187.25; maintenance and miscellaneous, $118.80; parking and tolls, $125; tires, $300; wash and wax, $75; insurance, $614.80; license, $34.90; and depreciation, $1780. What was his cost per mile last year? ______________

7. Meg Stone drove her car to and from work last year. Her records show a total of $1560 for fixed costs and $3740 for variable costs. If it cost Meg $0.36 per mile to drive, how many miles did she drive last year? ______________

8. You drove your car to and from work last year. Your records show a total of $1860 for fixed costs and $5000 for variable costs. If it cost you $0.40 per mile to drive, how many miles did you drive last year? ______________

Copyright © by Glencoe Division

Name ______________________ Date ______________

LESSON 8-6

Leasing an Automobile

When you lease an automobile you make monthly payments to the leasing agency. Your leasing costs include all the monthly payments, a security deposit, title fee, and license fee.

$$\text{TOTAL LEASE COSTS} = \left[\begin{matrix}\text{NUMBER OF}\\ \text{PAYMENTS}\end{matrix} \times \text{PAYMENT}\right] + \text{DEPOSIT} + \begin{matrix}\text{TITLE}\\ \text{FEE}\end{matrix} + \begin{matrix}\text{LICENSE}\\ \text{FEE}\end{matrix}$$

1. Teresa Spaulding leased a 4 × 4 pickup for use in her landscape business. She paid $120 a month for 60 months. She also paid a deposit of $1000, a title fee of $90, and a license fee of $125. What is the total least cost? ______

2. Jim Pavlov leased an Eagle Premire for $219.50 a month for 60 months. He paid a deposit of $250, a title fee of $95, and a license fee of $220. What is the total lease cost? ______

3. Rodney VinDevers leased a Chevy Corsica for business use as a salesman. The lease cost $199 a month for 48 months. He paid a deposit of $750, a title fee of $95, and license fee of $135. What is the total lease cost? ______

4. Tina Truett had an "open-end" lease for a Dodge Caravan for her home decorating business. The lease cost $315 a month for 60 months. She paid a deposit of $1200, a title fee of $135, and a license fee of $85. At the end of the lease, she can buy the van for its residual value of $3850.

a. What is the total least cost? ______

b. What is the total cost if she buys the van? ______

5. Doris Santoro had an "open-end" lease for a Chevy Lumina for her audio store. The lease cost $239 a month for 48 months. She paid a deposit of $550, a title fee of $65, and a license fee of $120. At the end of the lease, she can buy the car for its residual value of $4860.

a. What is the total lease cost? ______

b. What is the total cost if she buys the car? ______

6. Wong Go leased a Honda Accord for $199.88 a month for 48 months. He paid a deposit of $1515, a title fee of $95, and a license fee of $65. The lease carried a stipulation that there would be a $0.10 per mile charge for all miles over 60,000. He drove the car 67,640 miles. What is the total cost of leasing the automobile? ______

7. Bob Thayer leased a Jeep Wrangler for $209 a month for 60 months. He paid a deposit of $850, a title fee of $45, and a license fee of $60. The lease carried a stipulation that there would be a $0.12 per mile charge for all miles over 60,000. He drove the car 73,524 miles. What is the total cost of leasing the automobile? ______

Copyright © by Glencoe Division

Name ______________________________ Date ______________

LESSON 8-7

Renting an Automobile

When you rent an automobile, the total cost may include a daily rate, a rate per mile, gasoline charged to you, and insurance.

$$\text{COST PER MILE} = \frac{\text{TOTAL COST}}{\text{NUMBER OF MILES DRIVEN}}$$

---

1. Alicia Whitman rented a station wagon for her vacation. The wagon rented for \$25.40 a day plus 21¢ per mile. Alicia used the wagon for 7 days. She paid \$54.20 for gasoline during her 620-mile drive. What is Alicia's cost per mile to the nearest cent? ______________

2. The Neumeyers plan to fly to their vacation spot and then drive through the mountains. They arranged to rent a car for \$25.75 a day with no charge for mileage.

a. What will it cost the Neumeyers to rent the car for 10 days if they spend \$78 for gasoline and \$16.75 for miscellaneous fees? ______________

b. What will it cost per mile if they drive about 590 miles? ______________

3. The Stockton's are moving into a new house this weekend. To transport all their household belongings, they will need to rent a truck for one day. The rental cost will be \$45 a day plus 29¢ a mile. The collision waiver will cost \$25.

a. What will be their total cost if they pay \$15.50 for diesel fuel and drive 68 miles? ______________

b. What will be their cost per mile? ______________

4. Jerry Ritzenthaler rented a 14-foot truck to move into his new apartment. Jerry rented the truck for one day at a cost of \$37.50 per day plus 21¢ per mile. The collision waiver cost \$10.00. Gasoline cost \$44.60. Jerry drove 220 miles. What was his cost per mile? ______________

5. Lori Ann Cox flew to her vacation condominium and rented a sports car for 7 days. The car rents for \$129 per week with no charge for mileage. Find her cost per mile if gasoline cost \$31.64 and she drove 485 miles. ______________

6. You fly to a weekend business conference and rent a car for Friday, Saturday, and Sunday. You drive 115 miles and gasoline costs you \$7.50. The weekend rental rate on the car is \$49.50 plus 15¢ per mile. What is your cost per mile? ______________

7. Four of you decide to rent a limousine for the prom. The cost is \$79.95 per day plus \$0.38 per mile. Insurance costs \$75. You rent the limo for 2 days and drive 48 miles. The limo gets 8 miles per gallon. Gas costs \$1.23 per gallon.

a. What is the total cost of renting the limo? ______________

b. What is the cost per mile? ______________

c. What is the cost per person? ______________

Copyright © by Glencoe Division

Name ______________________ Date ______________

LESSONS 9-1, 9-2

Mortgage Loans, Total Interest

When you purchase a home, you will probably make a down payment and finance the remaining portion of the selling price with a mortgage loan from a bank or savings and loan association. A mortgage loan is usually repaid with interest in equal monthly payments. If you know the annual interest rate, the amount of the loan, and the length of the loan, you can use a table to find the monthly payment, the total amount paid, and the interest charged.

MORTGAGE LOAN AMOUNT = SELLING PRICE − DOWN PAYMENT

$$\text{MONTHLY PAYMENT} = \frac{\text{AMOUNT OF MORTGAGE}}{\$1000} \times \text{MONTHLY PAYMENT FOR A \$1000 LOAN}$$

AMOUNT PAID = MONTHLY PAYMENT × NUMBER OF PAYMENTS

TOTAL INTEREST CHARGED = AMOUNT PAID − MORTGAGE AMOUNT

1. Kung and So Lee offered $87,000 on a home that had been priced at $96,500. The seller agreed to the offer. A 20% down payment is required. What is the amount of the down payment? What is the amount of the mortgage loan needed to finance the purchase? ______ ______

MONTHLY PAYMENT FOR A $1000 LOAN			
Annual Interest Rate	Length of Loan (Years)		
	20	25	30
10.00%	$ 9.66	$ 9.09	$ 8.78
10.50%	9.99	9.45	9.15
11.00%	10.33	9.81	9.53
11.50%	10.67	10.17	9.91
12.00%	11.02	10.54	10.29
12.50%	11.37	10.91	10.68
13.00%	11.72	11.28	11.07
13.50%	12.08	11.66	11.46

2. Mary Cunningham offered $156,500 for a home that had been priced at $169,500. The seller agreed to the offer. A bank is willing to finance the purchase if she can make a down payment of 20%. What is the amount of her mortgage loan? ______

3. Danielle and Jim Bone have obtained a $70,000 mortgage loan at an annual interest rate of 11.50% for 30 years.

a. What is the monthly payment? ______

b. What is the amount paid? ______

c. What is the total interest? ______

4. Lee Hays has obtained a $96,000 mortgage loan at 12.50% interest for 25 years.

a. What is the monthly payment? ______

b. What is the amount paid? ______

c. What is the total interest? ______

5. How much can be saved in total interest by financing $60,000 at 13.00% for 20 years rather than 25 years? ______

6. How much can be saved in total interest by financing $60,000 at 13.00% for 25 years rather than 13.5% interest for 25 years? ______

Copyright © by Glencoe Division

Name ______________________ Date ______________

LESSON 9-3

Closing Cost

At the time you sign the documents to transfer ownership of your new home, you must pay any closing costs that the bank charges. Your closing costs may include fees for the bank's lawyers, credit checks and title searches, taxes, and the preparation of the documents.

CLOSING COSTS = SUM OF BANK FEES

Use the list of closing costs at the right to solve problems 1 and 2.

Credit report:	$ 35.00
Appraisal report:	$225.00
Title search:	$140.00
Survey:	$125.00
Recording & transfer fee:	$ 20.00
Legal fees:	$360.00
Loan origination fee:	2% of mortgage

1. Al and Viola Speer were granted a $55,000 mortgage. At the closing they will have to pay the closing costs shown plus real estate taxes of $825. What are the total closing costs? ______________

2. Pablo and Maria Rivera were granted $128,000 mortgage. At the closing they will have to pay the closing costs shown plus real estate taxes of $1920. What are the total closing costs? ______________

3. Joy and John MacAllister have agreed to purchase a house for $79,900. First National Savings & Loan is willing to lend the money at 10.75% for 25 years, provided the MacAllisters can make a $19,900 down payment. The total closing cost is 3.5% of the amount of the mortgage. What is their total closing cost? ______________

4. You are interested in purchasing a $144,000 home. You plan to make a 25% down payment and obtain a 12% mortgage for 20 years for the remaining amount through City Savings & Loan. Complete the form below in order to determine the total closing cost.

CITY SAVINGS AND LOAN ASSOCIATION
DISCLOSURE OF CLOSING COSTS

AMOUNT OF MORTGAGE: ______________ DATE: 3/1/—

Appraisal report	$255.00
Credit report	35.00
Loan origination fee: 2% of mortgage	______________
Recording costs	45.85
Survey and photos	245.60
Title search & insurance	60.00
Legal fees	325.00
Property taxes	475.25
Interest on the mortgage from 1st of the month to the closing date 3/16 (ordinary interest).	______________
Total	______________

Copyright © by Glencoe Division

Name ______________________________ Date ______________

LESSON 9-4

The Monthly Payment

Most mortgage loans are repaid in equal payments. Each payment includes an amount for payment of interest and an amount for payment of the principal of the loan. The amount of interest is calculated using the simple interest formula. Each payment you make decreases the amount of the principal you owe.

PRINCIPAL PAYMENT = MONTHLY PAYMENT − INTEREST PAYMENT

NEW PRINCIPAL = PREVIOUS PRINCIPAL − PRINCIPAL PAYMENT

Complete the table below.

	Mortgage Amount	Interest Rate	First Monthly Payment	Amount for Interest	Amount for Principal	New Principal
1.	$ 86,000	13%	$ 970.08	$931.67	$38.41	
2.	$165,000	12.5%	$1762.20			
3.	$ 42,500	10.00%	$ 410.55			

4. Julie and Barry Spinos purchased a house for $96,400. They made a 25% down payment and financed the remaining amount at 13% for 30 years. Their monthly payment is $800.36. How much of the first monthly payment is used to reduce the principal? ____________

5. Jim and Julie Speer purchased a home for $97,400. They made a down payment of $17,400 and financed the remaining amount at 11.00% for 25 years. Their monthly payment is $784.80. What is the new principal after the first monthly payment? ____________

6. The Harrises purchased a home for $87,300. They made a $31,500 down payment and financed the remaining amount at 12.00% for 30 years. Their monthly payment is $574.18.

a. How much of the first monthly payment is used to reduce the principal? ____________

b. What is the new principal after the first monthly payment? ____________

7. You purchase a home for $87,500. After a 20% down payment, you finance the remaining amount for 25 years at 11%. Your monthly payment is $686.70. Complete the repayment schedule below for the first 6 months of your loan.

	Monthly Payment	Amount for Interest	Amount for Principal	New Principal
a.				
b.				
c.				
d.				
e.				
f.				

Copyright © by Glencoe Division

Name ______________________________ Date ______________

LESSON 9-5

Real Estate Taxes

When you own a home, you will have to pay city or county real estate taxes. The amount of real estate tax that you pay in one year depends on the assessed value of your property and the tax rate. The assessed value is found by multiplying the market value of your property by the rate of assessment. Your tax rate may be expressed in mills per dollar of valuation. A mill is $0.001.

ASSESSED VALUE = RATE OF ASSESSMENT × MARKET VALUE

REAL ESTATE TAX = TAX RATE × ASSESSED VALUE

---

1. The Mariaskis' home is located in a community where the rate of assessment is 45% of the market value. The tax is $65 per $1000 of assessed value. The Schaffers' home has a market value of $97,400.

a. What is the assessed value? ______________

b. What is the yearly real estate tax? ______________

2. The rate of assessment in Fulton County is 35%. The tax rate is 81.31 mills. What is the real estate tax on a piece of property that has a market value of $38,500? ______________

3. Ron and Barbara Lugo live in a city where the tax rate is 83.21 mills. The rate of assessment is 30%. The property that the Lugos own has a market value of $367,500. What is their real estate tax for a year? ______________

4. Pater and Camilla Myers live in a home with a market value of $124,750. The rate of assessment is 40% and the tax rate is 112.8 mills. What is the Myers' real estate tax for a year? ______________

5. You live in a home in Bloom County. Your home has a market value of $80,000. Your rate of assessment is 45%. You pay total property taxes of 56.79 mills. Complete the form below to see how your property tax bill is distributed.

PURPOSE OF TAX	TAX RATE IN MILLS	REAL ESTATE TAX
County:		
County general fund	2.35	$__________
County parks	.50	$__________
Mental health levy	1.20	$__________
Gifted children's school	1.55	$__________
Transportation system	2.14	$__________
School:		
Local school	32.50	$__________
Vocational school	3.20	$__________
School building bonds	9.95	$__________
Others:		
Corporation—city	2.40	$__________
Police/fire pension	1.00	$__________
Total	__________	Total $__________

Copyright © by Glencoe Division

Name ______________________ Date ______________

LESSONS 9-6, 9-7

Homeowner's Insurance and Insurance Premium

When you own a home, you will probably purchase homeowner's insurance as protection against losses due to fire, theft, and personal liability. To receive full payment for any loss up to the amount of the policy, you must insure your home for at least 80% of its replacement value. Insurance companies use the amount of coverage on your home to calculate the amount of coverage on your garage, personal property, and for loss of use. The amount of your premium depends on the amount of insurance, the location of your property, and the type of construction of your home.

AMOUNT OF COVERAGE = PERCENT × AMOUNT OF COVERAGE ON HOME

Use the table at the right to answer problems 1-3.

Coverage	Percent of Coverage
Personal Property	50%
Loss of Use	20%
Garage and Other Structures	10%

1. The Kimbroughs' home has a replacement value of $82,700. They are insuring it for 80% of the replacement cost.

   a. What is the amount of insurance? ______

   b. What is the amount of coverage for personal property? ______

2. The Lugos' home has a replacement value of $287,000. It is insured for 90% of the replacement cost.

   a. What is the amount of insurance? ______

   b. What is the amount of coverage for loss of use? ______

3. The Ellisons have insured their home for $90,000. Their personal property coverage is 50% of the amount of their home coverage, personal liability coverage is 45%, and loss of use coverage is 20%.

   a. What is the amount of coverage for personal liability? ______

   b. What is the amount of coverage for personal property? ______

   c. What is the amount of coverage for loss of use? ______

Use the table at the right for problems 4 and 5.

4. Your brick home has a replacement value of $100,000 and is insured for 80% of the replacement value. You live in an area that has been designated fire protection class 3. Find the annual premium. ______

5. Your brick home has a replacement value of $75,000 and is insured for 80% of the replacement value. You live in an area that has been designated fire protection class 9. Find the annual premium. ______

ANNUAL PREMIUMS

Amount of Insurance Coverage	Brick/Masonry Veneer Fire Protection Class			
	1–6	7–8	9	10
$ 50,000	$137	$141	$185	$195
60,000	147	151	199	210
70,000	164	166	219	230
80,000	185	191	252	264
90,000	206	212	281	295
100,000	229	236	313	328
150,000	353	362	481	505

Copyright © by Glencoe Division

Name ______________________ Date ______________

LESSON 9-8

Other Housing Cost

In addition to your monthly mortgage payment, real estate taxes, and insurance payment, you will have other expenses for utilities, maintenance, and home improvements. The Federal Housing Administration (FHA) recommends that your total monthly housing cost be less than 35% of your monthly net pay.

1. Mario Orozco's monthly net pay is $3245. Housing expenses for November were:

Mortgage payment	$636.30
Real estate taxes	172.00
Insurance	29.50
Electricity	87.80
Telephone	28.40
Total Housing Cost	______

Is it within the FHA recommendation?

____

2. Kamil Saleb's monthly net pay is $2150. Housing expenses for March were:

Mortgage payment	$502.32
Insurance	27.50
Real estate taxes	130.91
Electricity	126.35
Telehphone	34.30
Water	23.60
Total Housing Cost	______

Is it within the FHA recommendation?

____

3. The Miquels had the following housing expenses for September: mortgage payment of $396.80, $34.15 for insurance premium, $139.40 for real estate taxes, $44.75 for home improvements, $51.20 for electricity, $29.75 for telephone service, $63.84 for natural gas, and $18.50 for water. Their monthly net pay is $2478.60.

a. What is their monthly housing cost? ______

b. For their monthly net pay, what is the recommended FHA maximum housing expenses? ______

c. Is their monthly housing cost within the FHA recommendation? ______

4. Lori and Mike Boyd have a combined monthly net income of $3395. Their records show that last year they paid $5484.60 in mortgage payments, $356 for insurance premiums, and $2240 in annual real estate taxes. They also had the annual expenses shown. Did they stay within the FHA recommendation?

Electricity	$1960.00
Water	194.50
Telephone	275.28
Washer/dryer	941.76
Painting	857.60
New carpeting	1231.75

____

5. You purchased a brick home for $162,500. You had a 20% down payment and financed the remainder at 10.00% for 25 years. Your property tax rate is 86.41 mills with an assessment rate of 25%. You had the housing expenses shown. How much net income do you need to be within the FHA recommendation?

Insurance	$545.00
Electricity	1140.37
Water	438.70
Telephone	507.96
Heating fuel	1297.74
Repairs	446.20

____

Copyright © by Glencoe Division

Name ______________________ Date ____________

Home Weatherization: A Simulation

Many homes have been constructed with insufficient insulation and without proper weatherization. Homes can be made more energy efficient by adding attic and crawl space insulation, by installing storm doors and storm windows, and by caulking windows and doors.

In order to conserve energy and save money on fuel bills, you have decided to investigate a home weatherization program for your house. To weatherize your home, you need to reduce the heat that is lost by conduction and infiltration. Heat lost by conduction is the heat that escapes through the building materials. All materials used in building construction reduce the flow of heat, but insulation if more effective in reducing the flow of heat. Heat loss by infiltration is heat lost through the leakage of air. Air can leak in and out of a house through the cracks around windows and doors, through foundation cracks, and around electrical outlets.

You have inspected your home and consulted an energy savings guide to come up with the data in the chart below.

ESTIMATE OF ANNUAL ENERGY SAVINGS

Type of Heat Loss	Proposed Changes to Building	Heat To Be Saved (Million Btu)
Infiltration	Caulk and weatherstrip all doors and windows	25.1
Conduction through: Floors	Insulate upper portion of basement walls	7.5
Ceilings	Add 6 in. of insulation to ceiling	15.0
Windows	Add storm windows	7.95
Walls	None—walls already insulated	0
TOTALS		55.55

In order to save energy, you decide to make the proposed changes to your home. Complete the chart below to estimate the cost of each change. Since you will do the work yourself, there will be no labor charge.

	Type of Material	Number of Units	Cost per Unit	Estimated Cost
1.	Caulking	10 tubes	5 tubes of $1.89 each 5 tubes at $4.29 each	
2.	Weather stripping	221 ft	17-ft roll costs $2.65	
3.	Foam panels for basement walls	992 sq ft	32-sq-ft panel costs $4.39	
4.	Insulation: 6-in. fiber glass batts	1250 sq ft	50-sq-ft roll costs $15.49	
5.	Storm windows	8	$135 per window	
			Total	

Copyright © by Glencoe Division

Name ______________________ Date ______________

Home Weatherization: A Simulation
(CONTINUED)

After estimating the cost of each energy-saving improvement, you want to determine the number of heating seasons it will take to recover the cost of the improvement. The pay-off time for one improvement is found by using this formula.

$$\text{PAY-OFF TIME} = \frac{\text{COST OF IMPROVEMENT}}{\text{MILLION BTU SAVED ANNUALLY} \times \text{COST PER MILLION BTU}}$$

Type of Heat	Cost per Million Btu
Natural gas	$ 9.50
Fuel oil	$14.20
Electricity	$32.80

Find the pay-off time for each improvement for each type of heating. Round your answers to the nearest hundredth.

	Improvement	Natural gas	Oil	Electricity
6.	Caulking and weather stripping			
7.	Basement panels			
8.	Ceiling insulation			
9.	Storm windows			

Find the pay-off time for all the improvements combined for each type of heating by substituting the total cost of improvements in the formula above.

10. Natural gas ______________

11. Oil ______________

12. Electricity ______________

Copyright © by Glencoe Division

Name ______________________________ Date ______________

LESSON 10-1

Health Insurance Premiums

An accident or sickness could cut off your income, wipe out your savings, and leave you in debt. To protect against overwhelming medical expenses, many people have health insurance. You can get health insurance by joining a group plan where you work. Your employer may pay part or all of the premium.

EMPLOYER'S CONTRIBUTION = TOTAL PREMIUM × EMPLOYER'S PERCENT

EMPLOYEE'S CONTRIBUTION = TOTAL PREMIUM − EMPLOYER'S CONTRIBUTION

1. Paul Woonan, a self-employed accountant, is married and enrolls in a non-group health insurance plan. The plan costs $2710 per year for family coverage. Paul chooses to pay the premiums monthly. What will he pay each month? ______

2. Donna Ray is employed at Blanco Manufacturing Company. Her total annual health insurance premium is $1976. Donna's employer pays 70% of the health insurance premium. How much does she pay per month for health insurance? ______

3. Joyce Timpe is employed by Allied Chemical Company. She has a family membership in the group comprehensive medical program. The annual premium includes $1940 for hospital insurance, $424 for surgical-medical insurance, and $128 for major medical insurance. Her employer pays 75% of the total cost. Her contribution is deducted from her weekly paycheck.

a. What is the total annual premium? ______

b. How much does the employer pay annually? ______

c. How much does Janet pay annually? ______

d. How much is deducted each week from her paycheck? ______

4. Oscar Ankebrandt is employed by Lion Department Store. He has a family membership in the group comprehensive medical and life insurance program. The annual premium includes $2585 for hospital insurance, $816 for surgical-medical insurance, $206 for major medical insurance, and $242 for group term life insurance. Lion pays for 55% of the comprehensive medical and 80% of the life insurance premium. How much is withheld from Oscar's biweekly pay for group comprehensive and life insurance? ______

5. You are employed by Bell Telephone Company, and you sign up for the vision care health insurance program at a cost of $1.25 per week. The insurance coverage includes $25 toward the cost of an eye examination and $120 toward the cost of one pair of glasses. You had your eyes examined, and the charge was $45. You bought new frames that cost $90 and lenses that cost $135.

a. How much did you pay in premiums for 2 years? ______

b. How much is the bill for your eye exam, lenses, and frames? ______

c. How much of this is covered by insurance? ______

d. How much do you pay? ______

Copyright © by Glencoe Division

Name ______________________________ Date ______________

LESSON 10-2

Health Insurance Benefits

A comprehensive health insurance plan includes hospital insurance, surgical-medical insurance, and major medical insurance. Most major medical policies have a deductible clause. A $250 deductible clause means that you must pay the first $250 of the amount not covered by your hospital and surgical-medical insurance. Most major medical policies also have a coinsurance clause. A typical one states that you must pay 20% of the amount remaining after you pay the deductible. Your insurance company pays the other 80%.

AMOUNT PAID BY PATIENT = DEDUCTIBLE − COINSURANCE AMOUNT

Complete the table below.

	Total Medical Bill	Amount Paid by Hospital and Surgical-Medical Insurance	Major Medical Deductible	Amount Subject to Coinsurance	Coinsurance Rate	Amount Paid by Patient
1.	$18,540	$16,290	$250		25%	
2.	$29,950	$21,870	$200		10%	
3.	$ 9160	$ 8666	$200		20%	
4.	$62,241	$56,450	$100		30%	

5. Paula Wiley had a total medical bill of $3260. Her hospital and surgical-medical insurance paid $2625. Her major medical has a $250 deductible and a 30% coinsurance clause. How much did Paula have to pay? ______________

6. Pierre Waterman's last hospital stay had a bill of $14,816.91. His hospital and surgical-medical insurance paid $12,302.70. His major medical has a $150 deductible and a 20% coinsurance clause. How much did Pierre have to pay? ______________

7. Rick Orland's comprehensive medical insurance includes hospital, surgical-medical, and major medical insurance. The major medical policy has a $200 deductible and 15% coinsurance clause. When Rick had surgery, his total bill was $17,540. The hospital and surgical-medical provisions of his insurance policy covered $14,032.

a. Find the amount not paid by the hospital and surgical-medical insurance. ______________

b. Find the coinsurance amount paid by Rick. ______________

c. Find the total amount paid by Rick. ______________

8. Your comprehensive medical insurance includes hospital, surgical-medical, and major medical insurance. The major medical policy has a $350 deductible and a 20% coinsurance clause. You had an emergency appendectomy, and your total bill was $10,640. The hospital and surgical-medical provisions of your insurance policy covered $9044. How much did you pay? ______________

Copyright © by Glencoe Division

Name ______________________ Date ____________

LESSONS 10-3, 10-4

Life Insurance — Term and Other

The main purpose of life insurance is to provide financial protection for your dependents in case of your death. You may purchase term life insurance, whole life insurance, limited payment life insurance, or endowment life insurance.

ANNUAL PREMIUM = NUMBER OF UNITS PURCHASED × PREMIUM PER $1000

Use the tables below to answer the problems.

ANNUAL PREMIUM PER $1000 OF LIFE INSURANCE: 5-YR TERM

Age	Male	Female
18	$ 2.64	$ 2.25
20	2.66	2.27
25	3.07	2.39
30	4.29	3.85
35	5.61	4.95
45	11.11	10.65
55	23.60	22.01
65	42.77	35.42

ANNUAL PREMIUMS PER $1000 LIFE INSURANCE

Age M	Age F	Term to Age 65	Whole Life	20-Year Payment	20-Year Endowment
20	25	$ 7.01	$13.61	$25.16	$45.22
25	30	7.87	15.67	27.63	45.49
30	35	9.03	18.30	30.60	46.04
35	40	10.66	21.74	33.97	47.42
40	45	12.93	26.25	38.69	49.13
45	50	15.88	32.15	44.49	52.56
50	55	20.27	39.92	51.38	57.23
55	60	28.43	49.49	60.15	63.00

Complete the table below.

	Insured	Sex	Age	Type of Insurance	Annual Premium per $1000	Coverage	Number of Units	Annual Premium
1.	Ricardo Oro	M	25	5-year term		$39,000		
2.	Doris Stelnick	F	35	Whole life		$25,000		
3.	Marilyn Wilson	F	45	20-year endowment		$28,000		
4.	Perry Zamora	M	35	20-year payment		$40,000		
5.	Paula Wyatt	F	55	Term to age 65		$24,000		

6. Calvin Moyer wants to purchase a $30,000 5-year term life insurance policy. He is 18 years old. What is his annual premium? ____________

7. Alicia Bitman, age 30, plans to purchase a $50,000 life insurance policy. She is considering a whole life policy. What is her annual premium? ____________

8. Ten years ago, you purchased a $70,000, 20-year payment life insurance policy. At that time, you were 30 years old.

	M	F
a. What is your annual premium?	____________	____________
b. How much have you paid in the past 10 years?	____________	____________
c. How much will you pay in the 20 years?	____________	____________

Copyright © by Glencoe Division

Name ______________________ Date ______________

LESSONS 10-5, 10-6

Certificates of Deposit and Effective Annual Yield

Your money earns interest at a higher rate when you buy a certificate of deposit than it does when you invest it in a regular savings account. Most certificates earn interest compounded daily. The annual yield is the rate at which your money earns simple interest in one year.

INTEREST EARNED = AMOUNT − ORIGINAL PRINCIPAL

$$\text{ANNUAL YIELD} = \frac{\text{INTEREST FOR ONE YEAR}}{\text{PRINCIPAL}}$$

Use the table below to answer the problems.

AMOUNT PER $1.00 INVESTED, DAILY COMPOUNDING

Annual Rate	Interest Period: 3 Months	One Year	2.5 Years	4 Years	6 Years	8 Years
5.75%	1.014278	1.059180	1.154458	1.258577	1.411952	1.584017
6.00%	1.014903	1.061831	1.161820	1.271224	1.433287	1.616011
6.25%	1.015529	1.064489	1.169103	1.283998	1.454945	1.648651
6.50%	1.016155	1.067153	1.176431	1.296900	1.476930	1.681950
6.75%	1.016782	1.069824	1.183806	1.309932	1.499246	1.715921
7.00%	1.017408	1.072501	1.191226	1.323094	1.521900	1.750579
7.25%	1.018036	1.075185	1.198693	1.336389	1.544896	1.785936
7.50%	1.018663	1.077876	1.206207	1.349817	1.568240	1.822006
7.75%	1.019291	1.080573	1.213768	1.363380	1.591936	1.858806
8.00%	1.019920	1.083278	1.221376	1.377079	1.615989	1.896348

1. Alicia Cox purchased a $2\frac{1}{2}$-year certificate of deposit for $15,000. The certificate earns interest at a rate of 7.00% compounded daily.

a. What is the amount of the certificate at maturity? ____________

b. What is the interest earned? ____________

2. Charles Demaize purchased an 8-year certificate of deposit for $45,000. The certificate earns interest at 8% compounded daily.

a. What is the certificate worth in 8 years? ____________

b. What has the certificate earned in those 8 years? ____________

3. The Trues have $30,000 they want to invest in a certificate of deposit. They can purchase a 4-year certificate that earns interest at a rate of 5.75% or a $2\frac{1}{2}$-year certificate that earns interest at a rate of 8.00%.

a. Which certificate offers a higher maturity value? ____________

b. What is the difference in the maturity values of the two certificates? ____________

4. What is the annual yield on a 1-year $15,000 certificate of deposit that earns interest at a rate of 7.50% compounded daily? ____________

5. You invest $25,000 in an 8-year certificate of deposit that earns interest at 8% compounded daily. What is the annual yield? ____________

6. Most banks pay higher rates of interest for longer periods of deposit. Why? ____________

Copyright © by Glencoe Division

Name ______________________ Date ______________

LESSONS 10-7, 10-8

Stocks and Stock Dividends

When you purchase a share of stock, you become a part owner of the company that issues the stock. The total amount you pay for the stock depends on the cost per share, the number of shares you purchase, and the stockbroker's commission. When you buy stock, you may receive dividends. If you consider annual dividends an important factor in investing, you can use the annual yield to compare different stocks as investments.

COST OF STOCK = NUMBER OF SHARES × COST PER SHARE

TOTAL PAID = COST OF STOCKS + COMMISSION

$$\text{ANNUAL YIELD} = \frac{\text{ANNUAL DIVIDEND PER SHARE}}{\text{COST PER SHARE}}$$

---

1. Emma Elliot purchased 400 shares of Zony Electronics at $19 per share and paid a 2% commission.

a. What was the cost of the stock? ______________

b. What was the total paid? ______________

2. Arthur McCrag purchased 320 shares of the Atlas Paper Company. He paid a 7.00% commission to the stockbroker. The stocks cost him $\$42\frac{3}{4}$ per share.

a. How much did the stock cost? ______________

b. What was the total paid? ______________

3. Alison Louis owns 90 shares of Allied Farm Chemicals for which she paid $49.60 per share. The company paid annual dividends of $4.08 per share. What is the annual yield? ______________

4. Arthur Delta owns 60 shares of Comp-Tronics for which he paid $3945.90. The $3945.90 includes a commission of $113.40. Comp-Tronics paid annual dividends of $7.56 per share. What is the annual yield? ______________

5. You are considering the purchase of either 100 shares of Data Control Corporation at $65.50 per share or 80 shares of Mini-Computer Limited at $74.25 per share. Data Control is expected to pay annual dividends of $6.40 per share, while Mini-Computer Limited is expected to pay $8.88 per share.

a. If you bought the 100 shares of Data Control Corporation, what would you receive in annual dividends? ______________

b. What would your annual yield be? ______________

c. If you bought 80 shares of Mini-Computer Limited, what would you receive in annual dividends? ______________

d. What would your annual yield be? ______________

6. Mario Gillespie owns 50 shares of Consolidated Industries for which he paid $92.625 per share and a 4% commission. Consolidated Industries paid annual dividends of $4.68 per share. What is the annual yield? ______________

Copyright © by Glencoe Division

Name ______________________________ Date ______________

LESSON 10-9

Selling Stocks

When you sell your stocks, the sale can result in either a profit or a loss. You profit when the sale amount minus the sales commission is greater than the original purchase amount. Your sale results in a loss when the sale amount minus the sales commission is less than the original purchase amount.

NET SALE = AMOUNT OF SALE − COMMISSION

PROFIT = NET SALE − TOTAL PAID

LOSS = TOTAL PAID − NET SALES

Round each answer to the nearest cent.

1. Adrain Engel owned 120 shares of Summit Oil Company stock, for which he paid a total of $8700. He sold the stock for $78 per share and paid a commission of $360.

a. What was the amount of the sale? ____________

b. What was the net sale? ____________

c. What was the profit or loss from the sale? ____________

2. Marty and Irene Benfeld purchased 300 shares of FCC International 2 years ago. They paid $7346 for the stock. Last week they sold the stock at $25.125 per share. The sales commission was 2.7%.

a. What was the amount of the sale? ____________

b. What was the net sale? ____________

c. What was the profit or loss from the sale? ____________

3. Jose Rodriguez recently sold 320 shares of stock for 44\frac{1}{4}$ per share. He paid a 4.3% sales commission. He purchased the stock for $9440.

a. What was the amount of the sale? ____________

b. What was the net sale? ____________

c. What was the profit or loss from the sale? ____________

4. Laura Long recently sold 45 shares of stock for $45 per share. She paid a 3% sales commission. She purchased the stock for $41.50 a share and paid a 3.6% commission.

a. What was the cost of the stock? ____________

b. What was the total paid? ____________

c. What was the amount of the sale? ____________

d. What was the net sale? ____________

e. What was the profit or loss from the sale? ____________

5. You have 850 shares of North Sea Oil stock. Recently a stockbroker offered you $74.90 per share for control of the stock. You purchased the stock for $54.20 per share. You paid a commission of 2.4% when you bought the stock. You will pay a sales commission of 3.2% when you sell. What would your profit or loss be if you sold now? ____________

Copyright © by Glencoe Division

Name ______________________ Date ____________

LESSON 10-10

Bonds

Bonds are issued by governments and large corporations to raise money. When you invest in bonds, you lend money to the corporation or government, and you are paid interest. When the bond matures, you will receive the face value of the bond.

ANNUAL INTEREST = INTEREST RATE × FACE VALUE

BOND COST = PERCENT × FACE VALUE

$$\text{ANNUAL YIELD} = \frac{\text{ANNUAL INTEREST}}{\text{BOND COST}}$$

Complete the table below. Round each percent to the nearest hundredth.

	Face Value of Bond	Quoted Price	Cost of Bond	Interest Rate	Annual Interest	Annual Yield
1.	$10,000	98		8%		
2.	$12,000	77		$7\frac{1}{2}$%		
3.	$ 5000	$85\frac{1}{2}$		$8\frac{1}{4}$%		
4.	$20,000	$91\frac{3}{8}$		$10\frac{1}{8}$%		
5.	$40,000	$85\frac{1}{4}$		12%		

6. Ralph Suarez purchases a $9500 bond at $88\frac{1}{4}$. It pays 6% annual interest.

a. What is the cost of the bond? ____________

b. What is the annual interest earned? ____________

c. What is the annual yield? ____________

7. Eva Rhodes purchases a $14,000 bond at $96\frac{3}{4}$. It pays $12\frac{5}{8}$% annual interest.

a. What is the cost of the bond? ____________

b. What is the annual interest earned? ____________

c. What is the annual yield? ____________

8. You use your savings to buy a $1500 bond at $82\frac{7}{8}$. It pays $8\frac{3}{4}$% annual interest.

a. What is the cost of the bond? ____________

b. What is the annual interest earned? ____________

c. What is the annual yield? ____________

Copyright © by Glencoe Division

Name ______________________ Date ______________

LESSON 11-1

Average Monthly Expenditure

You can manage your money better by keeping an accurate record of your expenditures. You will be able to evaluate your spending habits by keeping detailed records for a number of months.

$$\text{AVERAGE MONTHLY EXPENDITURE} = \frac{\text{SUM OF MONTHLY EXPENDITURES}}{\text{NUMBER OF MONTHS}}$$

Complete the table below.

	Name	MONTHLY EXPENDITURES				Average Monthly Expenditure
		July	August	September	October	
1.	P. Barker	$ 934.20	$1317.27	$1112.15	$1019.20	
2.	L. Lee	$ 823.40	$ 917.17	$1012.10	$ 987.43	
3.	K. Rue	$1134.14	$1572.99	$1996.61	$1475.50	
4.	T. Geariger	$1192.75	$1946.62	$1439.84	$1503.22	

5. In the table above, who had the highest average expenditure? ________

6. In the table above, who had the lowest average expenditure? ________

7. The Carsons' expenditures for this month are: rent, $412; groceries, $378; utilities, $219.55; gasoline, $50.50; entertainment, $54.80; medical bills, $62.40; and miscellaneous, $95.47. How much did they spend? ________

8. The Jacobs' expenditures for the past 7 months were: January, $1084.45; February, $886.40; March, $968.45; April, $1142.60; May, $995.80; June, $1379.86; and July, $1042.88. What is their average monthly expenditure? ________

9. Fay Teng's expenditures for the past 3 months were: August, $1735.50, September, $1829.42; and October, $1793.88. What is Fay's average monthly expenditure? ________

10. Jack Dodson has a budget of $990 a month. His budget includes rent, utilities, transportion, clothing, groceries, and miscellaneous expenses. For the past 6 months his expenditures were: January, $927.50; February, $984.40; March, $1032.32; April, $1195.55; May, $874.24; and June $943.37.

a. What is his average monthly expenditure? ________

b. How much over his budget is this? ________

11. You use the utility company's budget plan for paying your home heating bills. You pay $75 a month for 9 months. Your actual bills are: September, $28.95; October, $46.15; November, $68.46; December, $109.34; January, $115.26; February, $94.19; March, $75.66; April, $51.97; and May, $49.16. Compute the average for the 9 months. ________

Copyright © by Glencoe Division

Name ______________________ Date ______________

LESSONS 11-2

Preparing a Budget Sheet

A budget sheet outlines your total monthly expenses. It includes your living expenses, fixed expenses, and all annual expenses.

TOTAL MONTHLY EXPENSES = MONTHLY LIVING EXPENSES + MONTHLY FIXED EXPENSES + MONTHLY SHARE OF ANNUAL EXPENSES

Diane and Cory Legrand have a combined monthly net income of $1800. Use their budget sheet to answer the following questions.

1. What is the total of their monthly living expenses? ______

2. What is the total of their monthly fixed expenses? ______

3. What is the total of their annual expenses? ______

4. What is the monthly share of their annual expenses? ______

5. What is the total of their monthly expenses? ______

A MONEY MANAGER FOR ______________ DATE ______

MONTHLY LIVING EXPENSES		MONTHLY FIXED EXPENSES	
Food/Grocery Bill	$315.65	Rent/Mortgage Payment	$ 428.64
Household Expenses		Car Payment	$ 192.45
Electricity	$ 48.40	Other Installments	
Heating Fuel	$ 65.00	Appliances	$
Telephone	$ 40.50	Furniture	$
Water	$ 28.20	Regular Savings	$ 20.00
Garbage/Sewer Fee	$	EMERGENCY FUND	$ 40.00
Other	$	TOTAL	$
Transportation		**ANNUAL EXPENSES**	
Gasoline/Oil	$118.82	Life Insurance	$ 244.00
Parking	$ 32.00	Home Insurance	$ 357.00
Tolls	$ 11.00	Car Insurance	$ 306.60
Commuting	$	Real Estate Taxes	$ 1847.40
Other	$	Car Registration	$ 60.00
Personal Spending		Pledges/Contributions	$ 872.00
Clothing	$114.00	Other	$
Credit Payments	$	Total	$
Newspapers, gifts, etc	$ 27.27	MONTHLY SHARE *(Divide by 12)	$
Pocket Money	$ 25.00	**MONTHLY BALANCE SHEET**	
Entertainment		NET INCOME (Total Budget)	$
Movies/Theater	$ 32.00	Living Expenses:	$
Sporting Events	$	Fixed Expenses:	$
Recreation	$	Annual Expenses:	$
Dining Out	$ 57.00	TOTAL MONTHLY EXPENSES	$
TOTAL	$	BALANCE	$

6. Are the Legrands living within their monthly net income? ______

7. The Legrands receive pay raises that increase their net income by 5%. What is their new combined monthly net income? ______

8. Can the Legrands meet their total monthly expenses with their new monthly income? ______

9. If you have the same budget and the same net income as the Legrands had before they received pay increases, on what items would you try to reduce your spending in order to live within your monthly net income? ______

Copyright © by Glencoe Division

Name ______________________ Date ______________

LESSON 11-3

Using a Budget

You can use a budget to plan your future spendings. Using an expense summary, you can compare the amount spent with the amount you had budgeted. It is wise to include an emergency fund for unpredictable expenses, such as medical bills and repair bills.

PERCENT OF BUDGET = AMOUNT BUDGETED ÷ TOTAL BUDGET

Use the budget sheet on page 85 of your workbook to answer problems 1–6. Round each percent to the nearest tenth.

	Expenses	Amount Budgeted	Total Monthly Living Expenses	Percent of Budget	Percent for "Average Family"	Percent More or Less than Average
	Clothing	$114.00	$914.84	12.5%	6.8%	5.7% more
1.	Transportation		$914.84		20.3%	
2.	Groceries		$914.84		27.5%	
3.	Entertainment		$914.84		15.0%	
4.	Pocket money		$914.84		5.0%	

5. The Legrands increased their transportation expenses by $40 and decreased their groceries by $40. What percent of the budget for living expenses is the new transportation amount of $201.82? ____________

6. In June the Legrands' heating bill is reduced from $65.00 to $10.50.

a. How much do they save? ____________

b. By what percent are their household expenses reduced if their other expenses remain the same? ____________

7. Your monthly net income is $980. You allocate 10% for clothing, 27% for transportation, 30% for groceries, 8% for entertainment, 5% for pocket money, and the remainder for savings. Your actual expenses for 2 months are shown. What is the dollar amount allocated for each expense? Find the difference between the actual amount spent and the amount budgeted.

Item	Monthly Budget	Actual for Sept	Difference	Actual for Oct.	Difference	Two Months Combined
Clothing	________	$ 75.00	________	$136.27	________	________
Transportation	________	250.00	________	275.68	________	________
Groceries	________	324.50	________	275.61	________	________
Entertainment	________	121.60	________	65.32	________	________
Pocket Money	________	35.00	________	54.00	________	________
Savings	________	200.00	________	175.00	________	________
Totals	________	________	________	________	________	________

Are you spending over or under your budget? ____________

Copyright © by Glencoe Division

Name ____________________ Date ____________

LESSON 12-1

Hiring New Employees

To fill openings in your business, you can either recruit new employees or pay an employment agency to locate a candidate. The cost of recruiting may include advertising fees, interview expenses such as travel expenses, and hiring expenses such as moving expenses.

TOTAL RECRUITING COST = ADVERTISING EXPENSE + INTERVIEWING EXPENSE + HIRING EXPENSE

	1.	2.	3.	4.	5.
Position	**Statistician**	**Analyst**	**Instructor**	**Director**	**CPA**
Advertising Cost	$ 212.50	$ 350.50	$ 615.84	$ 68.40	$ 253.82
Interviewing Cost	56.80	762.20	8262.48	396.42	910.48
Hiring Cost	3952.00	5012.90	7467.97	2147.80	4697.80
Total Recruiting Cost					

6. Commerce Bank hired Pete Drexel as their new branch office manager, at an annual salary of $29,410.

Advertising Costs:	$421.27	
Interview Expenses:	P. Drexel	$85.25
	L. Lindsay	$94.60
Search Agency Fee:	15% of first year's annual salary	

What was the total cost of recruiting Pete Drexel for this position? ____________

7. In search of a new vice president, your company hired the Executive Placement Agency to locate candidates for the position. The agency's fee is 25% of the first year's salary, if you should hire one of their candidates. You have also run advertisements in professional magazines for a total cost of $849.04. You interviewed 3 people:

Amy Simpson	Chet Rucker	Denise Hoffman
Applied through the agency	Answered ad	Applied through the agency
Travel Cost: $490	Travel Cost: $104	Travel Cost: $229

You hired Denise Hoffman at an annual salary of $34,600. Your company paid her moving expenses of $1564.80, plus relocation expenses of $8100. What was your total recruiting cost? ____________

8. A major midwestern university hired Executive Search Consultants to assist in finding a new president. The consultant's fee is a flat $68,000. Advertisements cost $3160.96. The Search Committee brought 3 finalists to the campus for interviews.

	Tony Alfred	June Day	Scott Benham
Transportation	$410.25	$42.50	$694.30
Lodging and meals	118.80	197.46	212.40

The university named Tony Alfred their new president. Moving expenses of $914.40 plus relocation costs of $5500 were paid by the university. What was the total recruiting cost? ____________

Copyright © by Glencoe Division

Name ______________________________ Date ______________

LESSON 12-2

Administering Wages and Salaries

Your business may have a wage and salary scale for the positions in the company. A cost-of-living adjustment is a raise in your salary to help you keep up with inflation. A merit increase is a raise in your salary to reward you for the quality of your work.

NEW SALARY = PRESENT SALARY + COST-OF-LIVING ADJUSTMENT + MERIT INCREASE

	1.	2.	3.	4.
Job Title	**Clerk**	**Manager**	**Programmer**	**Accountant**
Present Salary	$11,780.00	$32,400.00	$22,500.00	$34,640.00
Cost-of-Living Increase	301.30	1,580.00	1,283.00	1,232.00
Merit Increase	452.60	2,517.00	658.47	1,232.00
New Salary				

5. Alan Russell is the credit manager at Value Line Shops. His annual salary is $20,040. Next month he will receive a 5.5% cost-of-living adjustment and a 4.0% merit increase. What will be his new annual salary? ____________

6. Sherry Spencer is the Senior Vice President for her company. Her annual salary is $76,425. She received a 6.4% merit increase this month. What is her new annual salary? ____________

7. Bert Zender will receive a promotion next month. He will receive a 12.5% salary increase, in addition to a 5.5% cost-of-living adjustment and a 6.8% merit increase. His present annual salary is $26,990. What will be his new annual salary? ____________

8. Wendy Tanner is the head computer analyst for Data Program Controls, Incorporated. She will receive a 5.1% cost-of-living adjustment in her salary this month. Her present annual salary is $42,600. What will be her new annual salary? ____________

9. After working at Bittman-Little for 8 months, you received your first performance review last week. As a result, you have been given a 7.5% merit increase based on your present annual salary of $21,640. What is your new salary? ____________

10. Killian Manufacturing awards merit increases based on performance and uses a sliding scale for cost-of-living adjustments. Employees at Killian who earn less than $18,000 receive a cost-of-living adjustment of 7.5%. Employees who earn $18,000 to $25,000 receive a cost-of-living adjustment of 6.5%. Employees who earn $25,000 or more receive a cost-of-living adjustment of 6.3%. Find the new annual salary for these 5 Killian employees. Their merit increases are given.

	Jim Spar	May Beggins	Ralph Stokes	Doug Precht	Janice Stites
Annual Salary	$18,000	$11,416	$31,860	$14,615	$27,916
Merit	6%	$3\frac{1}{2}$%	5.1%	0.7%	$6\frac{1}{4}$%
New Annual Salary					

Copyright © by Glencoe Division

Name ______________________________ Date ______________

LESSON 12-3

Employee Benefits

Employee benefits are offered in the form of various types of insurances, pension plans, and paid vacations. The total of your benefits is often figured as a percent of your annual gross pay.

$$\text{RATE OF BENEFITS} = \frac{\text{TOTAL BENEFITS}}{\text{ANNUAL GROSS PAY}}$$

Round to the nearest tenth percent.

Job Title	1. Administrator	2. Clerk	3. Programmer	4. Receptionist	5. Secretary	6. Vice President
Annual Gross Pay	$23,140	$18,600	$19,750	$12,200	$24,416	$92,412
Total Benefits	7,520	2,790	2,070	1,844	3,489	9,867
Rate of Benefits						

7. Brad Kahl is a computer processing supervisor. His annual salary is $32,190. Benefits consist of $1881.73 in vacation time, holidays worth $1287.60, health insurance premiums of $2300 paid by the employer, social security of $1995.78, medicare of $466.76, and unemployment insurance of $541.06.

a. What are the total benefits? ______

b. What is the rate of benefits? ______

8. Teri Schenk is employed as a supermarket cashier. Her annual salary is $17,470. Benefits consist of 1 week paid vacation, 8 paid holidays, 80% of a total health insurance package costing $2100, 3% unemployment insurance, 6.2% social security, and 1.45% medicare.

a. What are the total benefits? ______

b. What is the rate of benefits? ______

9. You work in a payroll office. Every January you complete the form below. Remember 6.2% of the first $62,700 of the annual salary is deducted for social security and 1.45% of all income is deducted for medicare.

Position	Annual Salary	Vacation	3% Health Ins.	3.6% Unemp. Ins.	6.2% Soc. Sec.	1.45% Med.	Total Benefits
Manager	$64,300	3 wk:					
Receptionist	18,600	2 wk:					
Clerk I	17,900	2 wk:					
Clerk II	15,400	2 wk:					

a. What is the rate of benefits for the manager? ______

b. What is the rate of benefits for the receptionist? ______

c. What is the rate of benefits for the clerk I? ______

d. What is the rate of benefits for the clerk II? ______

Copyright © by Glencoe Division

Name ______________________________ Date ______________

LESSON 12-4

Travel Expenses

If you travel for your business, you will be reimbursed for your authorized expenses. Travel expenses usually include transportation, lodging, and meals.

TOTAL TRAVEL EXPENSE = COST OF TRANSPORTATION + COST OF LODGING + COST OF MEALS + ADDITIONAL COST

	1.	2.	3.	4.	5.
Name	**H. Salkowski**	**O. Wines**	**S. Irelan**	**J. Jablonski**	**T. Mears**
Miles traveled	240.5	1980.90	410.50	1070.2	2540.6
Cost at $0.21/mile					
Meals	67.71	71.04	51.60	80.40	180.41
Lodging	98.60	168.40	64.90	148.10	260.80
Total Travel Expense					

6. Tom Maxwell is a consultant. This month his travel expenses include: airfare, $2690; taxi fares, $98.60; meals, $290; hotels, $590; miscellaneous, $38.00. What is Tom's total travel expense? ______________

7. Laura Kenman is the Superintendent for Fulham County. She travels to the 3 schools in her district every month. This month her travel expenses include: 246 miles traveled at $0.25 per mile; meals, $180.70; miscellaneous, $46.90. What is Laura's total travel expense this month? ______________

8. You went to Kansas City for a 3-day conference last week. Your travel expenses include the following: airfare, $220; taxi fare, $26.80; meals, $127.80; hotel, $89.90 per night for 2 nights; conference registration fee, $150. What is your total travel expense? ______________

9. Rose James, a systems engineer, spent 3 days at the Chicago Education Center doing bench marks on a bill-calc program. Her expenses included $289 airfare, $40 taxi fare plus 15% tip, $75 per night hotel for 2 nights, 2 dinners at $19.60 and $27.90, respectively, plus 20% tip in each case, and $12.50 for 2 breakfasts including tip. What is Rose's total travel expense? ______________

10. John Beck drove to a one-day business conference. He drove 224 miles round trip. His meals cost $21.90 plus 5.5% tax and 20% tip. Registration at the conference cost $55. He is reimbursed at $0.23 per mile plus gas. His car gets 28 miles per gallon and gas cost $1.24 per gallon. What is John's total travel expense? ______________

Copyright © by Glencoe Division

Name ______________________ Date ______________

LESSON 12-5

Employee Training

In-service training is often available to employees. Your company may grant you release time to attend these special classes. Release time allows you to attend the training sessions and still receive your regular wages while you are away.

TOTAL TRAINING COST = COST OF RELEASE TIME + COST OF INSTRUCTION + ADDITIONAL COST

	1.	2.	3.	4.
Purpose of Training	**Marketing New Products**	**Fuel Saving Tips**	**Management Training**	**New Phone System**
Number of Days	3	4	2	2
Daily Cost of Release Time	$486.00	$359.00	$ 794.00	$176.04
Daily Cost of Instruction	132.60	29.40	1240.00	150.00
Daily Cost of Supplies	78.40	77.90	48.96	87.90
Total Training Cost				

5. Two orthopedic surgeons attend a 3-day conference on surgery. Release time costs $890 per person per day, registration fee is $1060 per person, lodging and meals cost $175 per person per day, and supplies cost $194.60 per person. What is the total training cost? ______

6. The Holly Graphics Company is sending 4 of their designers to attend a 3-day workshop. The daily cost of release time is $173 per person. The registration fee is $68 per person. Each designer is also given a $50 per day miscellaneous allowance. What is the total training cost for the company? ______

7. You are going to a 7-day conference. The cost for release time will be $94.90 per day. The registration fee will be $580. Your travel expense for these 7 days will be $764.80. What will be the total cost for your training? ______

8. Six supervisors attend an in-service training session on better interpersonal communications. The 1-day session is conducted by a licensed psychologist at a cost of $670. Release time for 3 of the supervisors is $85 each; release time of the other 3 is $78.90 each. Supplies cost $21.50 per person. What is the total training cost? ______

9. Nine salespersons at Corner Furniture attended an 8-hour new products conference. The average hourly rate of pay for the 9 salespersons is $9.86. The sales manager, at an annual salary of $44,800, conducted the 1-day session after a 1-day prep period. Lunch was provided for the 9 salespersons and the sales manager at a cost of $11.80 per person. What is the total training cost? ______

Copyright © by Glencoe Division

Name ______________________________ Date ______________

Employee Benefits: A Simulation

Three companies have offered you a position with their business management departments. You are carefully debating among the offers, so that you can make a decision in your best interest.
Complete the benefits table. Round to the nearest cent and to the nearest tenth of a percent.

Company	Annual Wage	Vacation	Comp. Ins.	Unemp. Ins.	6.2% Soc. Sec.	1.45% Med.	Total Benefits	Rate of Benefits
Douglas	$18,800	2 wk:	3.0%	3.5%				
Bishop	$16,900	2 wk:	3.2%	3.2%				
Serba	$18,600	2 wk:	3.3%	3.8%				

Each company has its own cost-of-living adjustment and merit increase policies. You have made a comparison. Complete the table.

Company	Cost-of-Living Adjustment	Merit Increase
Douglas	5.0%	10.0%
Bishop	5.1%	11.6%
Serba	4.9%	12.1%

If you totaled the cost-of-living adjustment and merit increase, how much can you expect to earn after working one year with the company?

	Douglas	Bishop	Serba
Cost-of-Living Adjustment			
Merit Increase			
Annual Wage			
New Annual Wage			

FINAL THOUGHT

In selecting a job, should the starting salary be your only deciding factor? Why?

____

____

Copyright © by Glencoe Division

Name ______________________ Date ______________

LESSON 13-1

Manufacturing

The cost of manufacturing an item depends, in part, on the direct material cost and the direct labor cost.

PRIME COST PER ITEM = DIRECT MATERIAL COST PER ITEM + DIRECT LABOR COST PER ITEM

	Cost per Unit	Pieces per Unit	Direct Material Cost per Item	Labor Cost per Hour	Units per Hour	Direct Labor Cost per Item	Prime Cost
1.	$ 0.72	48		$ 8.50	340		
2.	$ 0.12	4		$13.60	170		
3.	$11.88	25		$10.40	270		
4.	$ 3.85	10		$36.90	28		

5. Pieces & Pieces Plastics manufactures plastic picture frames. Each sheet of plastic yields 184 frames. Each sheet costs $78.40. The direct labor charge is $9.60 per hour. The machine operator can cut 1 sheet per hour. What is the prime cost of manufacturing 1 frame? ______

6. A machine operator at Carson Can Company stamps labels on sheets of metal that are later made into cans. Each sheet can make 118 cans. The cost per sheet of metal is $10.60. The operator is paid $9.40 per hour and labels 120 sheets per hour. What is the prime cost of labeling 1 can? ______

7. Charlie Dakin, an industrial engineer, estimates the prime cost of manufacturing a staple remover. What is the prime cost per staple remover? ______

STAPLE REMOVER Cost Element	Cost per Item Material	Labor
Retractor	$0.017	$0.003
Wedge	0.005	0.002
Housing	0.005	0.014
Screw	0.005	0.003
Spring	0.004	0.002
Assembly	—	0.068
Tooling cost	0.002	0.053
Total		

8. Clarkson Cooking Products manufactures a garlic press. What is the prime cost of manufacturing 1,000,000 garlic presses? ______

GARLIC PRESS Cost Element	Cost per Item Material	Labor
Top handle	$0.099	$0.095
Bottom handle	0.079	0.007
Basket	0.034	0.019
Press plate	0.008	0.002
Sleeve	0.042	0.084
Assembly	0.038	0.195
Tooling cost	—	0.164
Total		

Copyright © by Glencoe Division

Name ______________________ Date ______________

LESSON 13-2

Break-even Analysis

A break-even analysis tells you how many units of a product must be made and sold to cover the cost of production. Any item sold beyond the break-even point results in a profit for your business.

$$\text{BREAK-EVEN POINT IN UNITS} = \frac{\text{TOTAL FIXED COSTS}}{\text{SELLING PRICE PER UNIT} - \text{VARIABLE COSTS PER UNIT}}$$

1. The Johnstowne Company manufactures porcelain figurines. The fixed costs for the product total $124,362.88. The average selling price per figurine is $210.10. The variable cost per bowl is $54.65. What is the break-even point in number of figurines? ______

2. Whisk-Off dog collar is produced by the Pet Mart Company. They have total fixed costs of $12,000,000 in the manufacture of the collar. The selling price of each collar is $6.95. The variable cost per collar is $4.55. What is the break-even point in number of collars ______

3. The Tobby Toy Company manufactures teddy bears. They have total fixed costs of $232,548.996 in the production of the bears. The selling price of each bear is $19.16. The variable cost per bear is $15.15. What is the break-even point in number of bears? ______

4. Your company produces In-A-Stick glue in 8-oz tubes. The total fixed costs for the production of the glue is $477,999.50. The variable cost per tube is $0.38, while the selling price is $1.16 per tube. What is the break-even point in tubes of In-A-Stick glue? ______

5. Forever Green Plastics manufacturers garden tool sets for cultivating. Total fixed costs are estimated at $64,800. The variable cost per set is $2.95 and the selling price is $8.60. What is the break-even point? ______

6. Fleckston Rubber Company manufactures hood release levers for an automobile manufacturer. The fixed costs are $53,090. Each lever is sold for $3.76. The variable cost per lever is $1.77. What is the break-even point? ______

Copyright © by Glencoe Division

Name ______________________________ Date ______________

LESSON 13-3

Quality Control

You can control the quality of items being mass produced by using a quality control chart. If too many items are defective, the situation is "out of control."

$$\text{PERCENT DEFECTIVE} = \frac{\text{NUMBER DEFECTIVE}}{\text{TOTAL NUMBER CHECKED}}$$

1. Rita DeVane, a quality control inspector for Berlin Sock Company, checked a sample of 875 socks. She found 94 defective samples. If more than 7% of the sample inspected is defective, the process is out of control. What percent of the sample is defective? Is the process in or out of control? ______________

2. Larry Watson checks 400 samples per hour of calculators produced by the Sherman Company. The production process is in control if 1% or less of the samples is defective. At 6:00 A.M. Larry found 3 of 400 samples defective. At 7:00 A.M. he found 6. Then at 8:00 A.M. he found 1. What percent of each inspection was defective? Was the process in or out of control each hour?

6 A.M. ______________

7 A.M. ______________

8 A.M. ______________

3. Quality control inspectors at Ireland Glass Works recorded the following data during one 8-hour shift. Compute the percent defective. Record the information on the quality control chart at the bottom of the page.

ITEM: GLASS GOBLETS #131								
TIME:	8 A.M.	9 A.M.	10 A.M.	11 A.M.	12	1 P.M.	2 P.M.	3 P.M.
Number Checked	25	40	20	25	25	20	40	25
Number Defective	1	1	2	1	1	1	3	1
Percent Defective	____	____	____	____	____	____	____	____

4. You work as a quality control inspector at Ireland Glass Works. You recorded the data for the midnight to 7 A.M. shift. Compute the percent defective. Record the information on the quality control chart at the bottom of the page.

ITEM: GLASS GOBLETS #131								
TIME:	12	1 A.M.	2 A.M.	3 A.M.	4 A.M.	5 A.M.	6 A.M.	7 A.M.
Number Checked	40	25	40	20	25	40	40	20
Number Defective	2	1	1	1	2	2	1	2
Percent Defective	____	____	____	____	____	____	____	____

3.

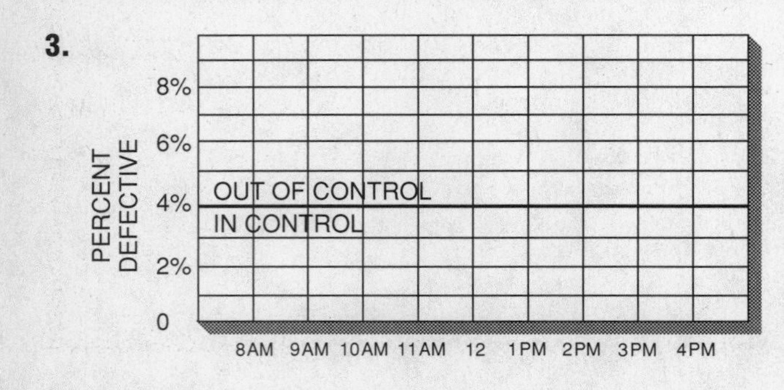

4.

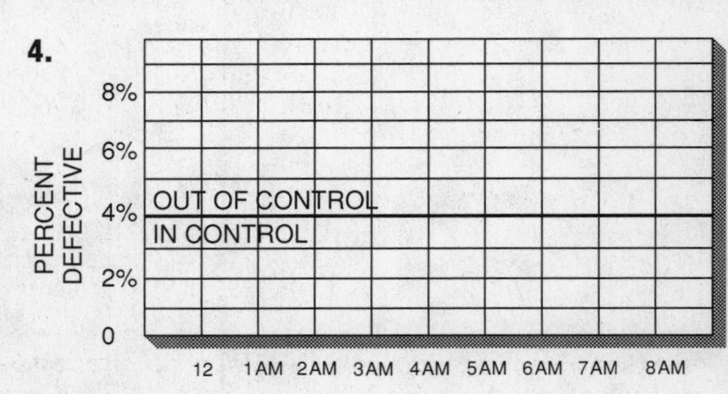

Copyright © by Glencoe Division

Name ______________________________ Date ______________

LESSONS 13-4, 13-5

Time-Study — Percent Allocation of Time

A time-study is used to determine the average time an employee takes to do a job or a variety of related activities. You can use the results of a time-study to plan how much time to allow for certain job activities.

$$\text{NUMBER OF UNITS} = \frac{\text{ACTUAL TIME WORKED}}{\text{AVERAGE TIME REQUIRED PER UNIT}}$$

$$\text{PERCENT OF TIME SPENT ON ACTIVITY} = \frac{\text{TIME SPENT ON ACTIVITY}}{\text{TOTAL TIME}}$$

1. Eric Williams prepared a time-study to determine the average time required to process passengers through the U.S. Customs at the new international airport.

a. What is the average time required for each task, to the nearest hundredth of a minute?

b. How many passengers can be inspected by 1 customs officer in 1 hour?

____

c. What percent of the inspector's time is spent on completing required documents/forms?

____

Task	OBSERVATIONS IN MINUTES #1	#2	#3	#4	#5	Average Time
Open suitcases	0.6	0.7	0.5	0.6	0.3	______
Inspect suitcases	3.0	2.9	3.2	3.9	2.8	______
Close suitcases	1.1	1.0	1.1	1.3	1.2	______
Complete required documents/forms	2.5	2.1	3.1	3.1	2.5	______
					Total	______

2. The Forest Valley Schools prepared a time-study of Paula Hughs' job as manager of a computer lab. What is the average time required for each task? What is the average number of hours worked each day? With respect to the average time, what percent of Paula's time is spent on each task?

Task	TIME IN HOURS M	T	W	TH	F	Average Time	Percent of Total
Helping students	4.0	3.9	4.1	4.6	4.0	______	______
Helping faculty	0.2	0.5	0.6	0.5	0.6	______	______
Group instruction	2.1	2.1	3.2	2.0	1.3	______	______
Paperwork	0.7	0.7	0.6	0.3	0.9	______	______
Reviewing programs	0.6	1.2	0.9	0.8	2.0	______	______
Coffee breaks	0.3	0.1	0.4	0.3	0.6	______	______
Totals	______	______	______	______	______	______	______

Copyright © by Glencoe Division

Name ______________________________ Date ______________

LESSON 13-6

Packaging

Putting your merchandise in containers for shipment is the last step in the production process. The size of the package depends on the size of the finished product.

DIMENSIONS: LENGTH, WIDTH, HEIGHT

---

1. Barry Thigpen packs books for a local warehouse. A new edition of the Heritage Dictionary is ready to be boxed for shipment. Each dictionary measures 11 in. × 8.5 in. × 4 in. The carton for these dictionaries is $\frac{1}{8}$-inch thick. What are the dimensions of the cartons if the dictionaries are laid flat, in 2 stacks of 6 dictionaries? ______________

What are the dimensions of the cartons if the dictionaries are arranged in only 1 stack of 8 dictionaries, also laid flat? ______________

2. Action batteries are packed 36 to a carton. The carton is made of 0.6 cm thick cardboard. There are no spacers. One battery measures 10 cm × 25 cm × 43 cm. What are the dimensions of the cartons if the batteries are arranged in 4 stacks of 9 batteries? ______________

What are the dimensions of the cartons if the batteries are arranged in 2 stacks of 18 batteries? ______________

3. You work in a crystal factory. A client has ordered 240 goblets. You have to pack each goblet in individual boxes before placing them in containers for shipment. The measurement of each box is 3 in. × 6 in. × 6 in. The shipping container is made of cardboard $\frac{1}{8}$-inch thick. No spacers are needed since each individual box is filled with foam shock absorbers. What are the dimensions of the containers if the boxes are arranged in 3 stacks of 10 boxes? ______________

How many containers would you need all together for this shipment? ______________

4. A popular beverage is sold in bottles with dimensions as shown. The bottles are packed 12 to a case for shipping (2 rows with 6 bottles per row). The case is made of $\frac{1}{8}$-inch thick corrugated cardboard with $\frac{1}{16}$-inch cardboard dividers. An extra layer of $\frac{1}{8}$-inch cardboard is placed on the bottom for a cushion. What are the dimensions of the case?

____

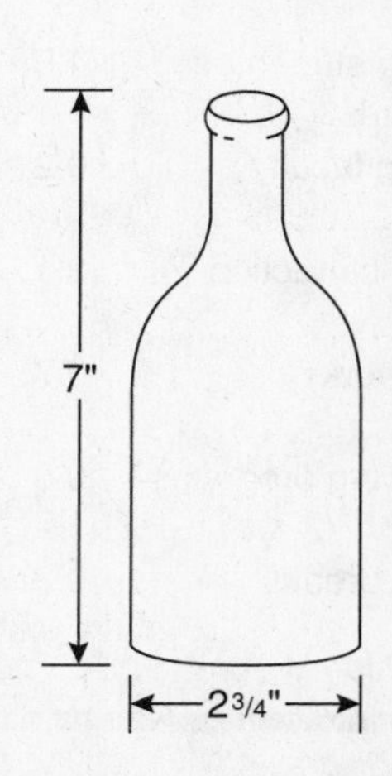

Copyright © by Glencoe Division

Name ______________________________ Date ______________

Manufacturing: A Simulation

You are the production manager at R&J Plastics Company. The company added two new items to the production line six months ago. Large sheets of plastic are used to make these items. You have been asked to update the prime cost of manufacturing the new product. Round to the nearest cent.

Item	Cost per Sheet	Pieces per Sheet	Direct Material Cost per Piece	Labor Cost per Hour	Pieces per Hour	Direct Labor Cost per Piece	Prime Cost
Picture Frame	$118.50	150		$10.80	257		
Paper Clip Dispenser	$145.96	178		$11.60	504		

R&J has decided to raise the selling price. How many units of each product must be made and sold to cover the production expenses at the new price?

	Picture Frame	Paper Clip Dispenser
Total Fixed Costs	$15,467.00	$13,612.00
Variable Cost per Unit	$1.41	$1.09
New Selling Price per Unit	$2.92	$2.05
Break-even Point in Units		

An inspection of 200 units of each item is made every 2 hours. If more than 6% of the sample is defective, the process is out of control. Complete the quality control chart for the picture frame. Round to the nearest percent.

Time	Number Checked	Number Defective	Percent Defective	In or Out of Control
8:00 A.M.	200	10		
10:00 A.M.	200	13		
12:00 NOON	200	15		

FINAL THOUGHT

What are two ways R&J Plastics can keep its production costs down?

____

____

Copyright © by Glencoe Division

Name ______________________________ Date ______________

LESSONS 14-1, 14-2

Trade Discounts and Complement Method

When you buy from a supplier, you usually get a trade discount. This is a markdown from the list price, or catalogue price, and represents a savings for your business. The net price is the amount you actually pay for the item. You can find the net price directly by using the complement of the trade-discount rate.

TRADE DISCOUNT = TRADE-DISCOUNT RATE × LIST PRICE

NET PRICE = LIST PRICE − TRADE DISCOUNT

NET PRICE = COMPLEMENT OF TRADE-DISCOUNT RATE × LIST PRICE

Round all answers to the nearest cent.

	1.	2.	3.	4.	5.	6.
List Price	$810.00	$61.40	$510	$310.61	$24.68	$6318.14
Trade-discount Rate	25%	40%	20%	15%	3%	$25\frac{1}{4}$%
Trade Discount						
Net Price						

Use the complement method.

7. Craig Milton needs to order paper for his company. In the Spencer catalogue a 20% trade discount is offered for a price of $481.60. What is the net price? ______________

8. The Bennetts own a craft store. They place an order for 100% wool yarn with a supplier who offers them a 17% trade-discount rate. The total list price for the yarn is $1016.14. How much do the Bennetts have to pay? ______________

9. Your garage is ordering radial tires for compact cars. The Union Tire Company offers you a 15% trade-discount rate for 56 tires. The list price for these tires is $61.60 each. Tire Trade has the same tire with as 12.5% trade-discount rate. The list price for Tire Trade is $59.90 each. Which supplier is less expensive? ______________

How much would you save if you purchased the 56 tires from the less expensive supplier? ______________

10. Fitzers Hardware receives a 25% trade discount from the wholesale hardware supply house. Find the total list price and net price on this invoice.

	Number	Description	Unit Quantity	Unit List Price	Total List Price	Net Price
a.	WH147X	Claw hammer	10	$12.98	$ ______	$ ______
b.	WH271B	Paper towel holder	20	13.95	______	______
c.	WH718XM	Metric bits	15	11.98	______	______
d.	WH791ZE	English bits	25	11.98	______	______
e.	WH918D	Drill	7	36.98	______	______

Copyright © by Glencoe Division

Name ____________________ Date ____________

LESSON 14-3

Trade-discount Rate

If you know the list price and the net price of certain items, you can calculate the trade-discount rate.

TRADE DISCOUNT = LIST PRICE − NET PRICE

$$\text{TRADE-DISCOUNT RATE} = \frac{\text{TRADE DISCOUNT}}{\text{LIST PRICE}}$$

Round to the nearest tenth of a percent.

	1.	2.	3.	4.	5.	6.
List Price	$210.00	$21.90	$18,241.00	$164.00	$500.00	$8143.25
Net Price	$187.95	$18.62	$14,592.80	$136.94	$300.00	$5252.40
Trade Discount						
Trade-discount Rate						

7. Renee Oswald ordered 10-speed bicycles for her store. The list price was $310 each. She paid $271.25 each. What was the trade-discount rate? ____________

8. Michael Reiley purchased 12 black walnut tables for his furniture store. The list price for the tables was $2410 each. Alfred paid $2120.80. What was the trade-discount rate? ____________

9. The Bolton School ordered new textbooks for the clerical recordkeeping class. The publisher's list price for the total was $1425. The school paid $1111.50. How much of a trade-discount rate was offered to the school? ____________

10. Trendil's is offering a trade discount to their clients if they purchase 10 or more Tiffany-style lamps. The manufacturer's list price is $136.98 per lamp. Trendil's is selling them for $104.10 each. How much is the discount? ____________

What is the trade-discount rate? ____________

11. You are the purchasing agent for a major engineering firm. You have just ordered 30 new computer components. The manufacturer's list price is $791.60 each. You negotiated a net price of $587.76 each. How much of a trade-discount rate did you negotiate? ____________

12. If you divide the net price by the list price and take the complement of your answer, you will have the trade-discount rate. Try it on these.

	List	Net	Net ÷ List	Trade-discount Rate
a.	$ 90.00	$ 62.55	________	________
b.	164.80	137.61	________	________
c.	294.12	229.42	________	________

Copyright © by Glencoe Division

Name ______________________________ Date ______________

LESSONS 14-4, 14-5

Chain Discounts and Complement Method

Chain discounts are offered to encourage you to place a larger order. A chain discount is a series of individual discounts. Complements may be used to find the net price directly or the single equivalent discount (SED).

NET-PRICE RATE = PRODUCT OF COMPLEMENTS OF CHAIN-DISCOUNT RATES

NET PRICE = NET-PRICE RATE × LIST PRICE

SED = COMPLEMENT OF NET-PRICE RATE

DISCOUNT = SED × LIST PRICE

Solve to the nearest cent.

	List Price	Chain Discount	First Discount	First Net Price	Second Discount	Second Net Price
1.	$1200.42	16% less 5%				
2.	$ 116.81	30% less 5%				
3.	$8492.80	20% less 7%				
4.	$ 312.90	12% less 8%				

Use the complement method.

5. Midtowne Furniture Center offers trade discounts and additional discounts to encourage large orders. What is the net price per item if each invoice total is high enough to obtain the additional discount?

	Item	List	Quantity	Trade Discount	Additional Discount	Net Price
a.	Oak table	$365.74	15	15%	8%	______
b.	Brass bed	749.80	7	22%	4%	______
c.	Sofa	514.60	10	18%	5%	______

6. Find the net price and SED (nearest tenth percent) for each of these auto parts.

	Part Number	Suggested List Price	Chain Discounts	Net Price	SED
a.	XL461	$ 68.40	30% less 10% less 25%	______	______
b.	ZW912	116.80	30% less 15% less 10%	______	______
c.	RL912	391.96	20% less 20% less 5%	______	______
d.	TK618	8.41	25% less 35% less 10%	______	______

Copyright © by Glencoe Division

Name ______________________________ Date ______________

LESSONS 14-6, 14-7

Cash Discounts—Ordinary Dating and EOM Dating

If you pay your invoice within a specified number of days, the supplier may offer you a cash discount based on the net price. You then pay the cash price. In ordinary dating, the discount term is from the date of the invoice. In EOM dating, the discount term is from the end of the month.

CASH DISCOUNT = CASH-DISCOUNT RATE × NET PRICE

CASH PRICE = NET PRICE − CASH DISCOUNT

---

1. The net price of goods from Yearling Tree Store to the Ray-Smyth law office was $325.09 for indoor plants. The invoice was paid in 7 days. The terms on the invoice were 2/10, net/30. What was the cash discount? ______

What was the cash price? ______

2. Pearson Farms sells fresh produce to local markets. They offer a 4/5, n/10 discount to all their clients. Georgie's Market just placed an order for assorted vegetables. The net price was $816.12. How much can Pearson Farms expect to receive, if Georgie's Market paid one day after the date on the invoice? ______

3. An invoice from Overhead Doors, Inc., to Safeway Lumber Company had a net price of $12,162.04. The terms were 2/10 EOM. The invoice was dated April 20 and was paid May 9. What was the cash price? ______

4. An invoice from Egede Goods to Bowers Wholesale Grocers carried a net price of $9618.40. The terms were 5/10 EOM. The invoice was dated August 20 and was paid September 12.

a. What was the cash price? ______

b. What did it cost Bowers to pay late? ______

5. Jodee's Department Store received this invoice from the Shoe Gallery.

Date	Invoice #	Account #	Store #	Terms	Vendor #
3/20	81624	84 102	126	1/20 EOM	212

Customer Order #	Quantity	Unit Price	Amount
812	15	$ 58.98	______
916	10	102.00	______
981	25	76.40	______
		Total	______

a. Fill in the amount column.

b. What is the last date that Jodee's can take the cash discount? ______

c. What is the cash price of the invoice if paid by 4/8? ______

Copyright © by Glencoe Division

Name ______________________ Date ____________

LESSONS 15-1, 15-2

Markup and Markup Rate

When your business sells an item, it is at a higher price than the original cost of the item. The difference is the markup. This markup can also be expressed as a percent of the selling price.

MARKUP = SELLING PRICE − COST

$$\text{MARKUP RATE} = \frac{\text{MARKUP}}{\text{SELLING PRICE}}$$

Kathryn Howard is a clerk at Paint Valley. She computes the markup and markup rate (to the nearest percent) on the selling price for these items.

	Item	Cost	Selling Price	Markup	Markup Rate on Selling Price
1.	Spray paint, 12 oz.	$3.16	$ 4.61	______	______
2.	Linseed oil, 1 gal.	9.29	12.26	______	______
3.	Trim brush, 1.5 in.	1.94	2.50	______	______
4.	Drop cloth, 9 in. × 12 in.	4.70	5.64	______	______
5.	Paint remover, 1 qt.	2.39	4.25	______	______

6. Seymour Cycle Shop buys new bicycles for $289.40. The bikes are sold for $482.00. What is the markup? ______

What is the markup rate as a percent of the selling price? ______

7. Allen Home Improvement Center buys kerosene heaters for $119.68 each. The retail selling price is $209.99. What is the markup? ______

What is the rate of markup? ______

8. American Merchandising distributes pharmaceutical products to various retail outlets. It costs each outlet $2.04 for a 4-ounce bottle of cough syrup. Determine the markup and markup rate as a percent of the selling price for each outlet.

Outlet	Selling Price	Markup	Markup Rate
Foodland Pharmacy	$3.86	______	______
Mainline Pharmacy	2.99	______	______
Phil's Pharmacy	2.89	______	______

9. You work at your local supermarket in the dairy department. Compute the cost per item, markup, and markup rate as a percent of the selling price.

Item	Case Cost	Number per Case	Cost per Item	Selling Price	Markup	Markup Rate
Cream cheese	$28.36	36	______	$1.39	______	______
Colby cheese, 12 oz.	36.85	24	______	3.49	______	______
American cheese, 1 lb.	25.70	24	______	1.94	______	______

Copyright © by Glencoe Division

Name ______________________ Date ______________

LESSONS 15-3, 15-4

Net Profit and Net Rate

The markup on the products that you sell must cover your overhead or operating expenses. When the markup of an item is greater than its overhead expenses, you make a profit on the item.

NET PROFIT = MARKUP − OVERHEAD

$$\text{NET-PROFIT RATE} = \frac{\text{NET PROFIT}}{\text{SELLING PRICE}}$$

Round to the nearest whole percent.

	Item	Selling Price	Cost	Markup	Overhead	Net Profit	Net-profit Rate
1.	Batteries	$ 2.99	$ 1.69	________	$ 0.15	________	________
2.	Stereo	1090.00	745.50	________	131.50	________	________
3.	Paper Towels	0.89	0.29	________	0.03	________	________
4.	Dress	189.99	58.60	________	21.40	________	________

5. PhotoMart purchased some film for $1.29 per roll. The selling price is $4.12 per roll. The operating expenses are estimated to be 25% of the selling price. What is the net profit per roll? ________

What is the net-profit rate per roll? ________

6. Down's Linen Store is selling white cotton sheet sets for $35.96. An estimated 25% is included in the selling price to cover overhead. The original price of the bedspread was $18.40. What is the net profit for each bedspread? ________

What is the net-profit rate? ________

7. The Sneaker Shoe Shop buys basketball shoes at $54.66 per pair. The selling price is $114.70. Approximately 20% of the selling price pays for overhead. What is the net profit per pair? ________

What is the net-profit rate? ________

8. You manage a local convenience store. Included in the selling price of every item sold is 15% to pay for overhead. What is the net profit for each item listed below? What is the net-profit rate?

Item	Selling Price	Cost	Markup	Overhead	Net Profit	Net-profit Rate
Rice	$1.49	$0.76	________	________	________	________
Carrots	1.19	0.88	________	________	________	________
Coffee	4.99	3.89	________	________	________	________
Potato chips	2.19	1.49	________	________	________	________

Copyright © by Glencoe Division

Name ______________________ Date ______________

LESSON 15-5

Determining Selling Price — Markup Based on Selling Price

You can use the cost of an item and the desired markup rate based on selling price to figure the best selling price for your products.

$$\text{SELLING PRICE} = \frac{\text{COST}}{\text{COMPLEMENT OF MARKUP RATE}}$$

Round to the nearest cent.

	1.	2.	3.	4.	5.	6.
Cost	$75.49	$12.00	$172.00	$89.61	$42.99	$18.50
Markup Rate (selling price base)	42%	50%	12%	30%	60%	28%
Complement of Markup Rate						
Selling Price						

7. Francis Furniture sells imported rosewood dining tables. The cost for one model is $762.87. Francis Furniture sells the table at a 15% markup based on the selling price. What is the selling price? __________

8. Universal Sounds and Records sells compact discs at a markup that is 42% of the selling price. If one CD costs Universal $5.90 to purchase, what is the selling price? __________

9. Binder's Bookstore purchases paperback novels from Parker Publishing Company. The markup for each book is 56% of its selling price. Binder's pays $2.90 for each book. What is the selling price? __________

10. Your company designs small silk-screen prints for $35 each. You sell them to a chain department store at a 25% markup based on selling price. They sell the screens to the consumers at a 40% markup based on selling price. What is the selling price of one screen to the department store? __________

What is the selling price of one print to the consumer? __________

11. It costs your company $8.89 to manufacture a steering component for a car. You sell the component to a distributor for $10.99. The distributor sells the component to auto parts stores at a 21% markup based on selling price. Auto parts stores sell the components to consumers at a 60% markup based on selling price. What is the selling price to the consumer? __________

Copyright © by Glencoe Division

Name ______________________ Date ______________

LESSONS 15-6, 15-7

Determining Selling Price — Markup Based on Cost

You can figure out the markup rate by comparing the markup to the original cost of the merchandise. You can use the cost of an item and the desired markup rate based on cost to figure the selling price of the item.

$$\text{MARKUP RATE} = \frac{\text{MARKUP}}{\text{COST}}$$

$$\text{MARKUP} = \text{MARKUP RATE} \times \text{COST}$$

$$\text{SELLING PRICE} = \text{COST} + \text{MARKUP}$$

Lindsay Carlson is a clerk at Hi-Gloss Paint. She computes the markup and markup rate (to the nearest percent) on the cost for this latest shipment of items.

	Item	Cost	Selling Price	Markup	Markup Rate Based on Cost
1.	Spray paint, 14 oz	$ 3.98	$ 6.57	______	______
2.	Linseed oil, 1.5 gal.	12.40	21.20	______	______
3.	Trim brush, 1.75 in.	1.01	1.61	______	______

4. Jane Freem buys jeans for a department store. She pays $21.26 per pair. The store sells each pair for $42.50. What is the markup rate based on cost? ______

What is the markup? ______

5. Happy Home Store buys towels from the manufacturer for $38.16 a dozen. The towels are sold at $9.65 per towel. What is the markup rate based on cost for one towel? ______

What is the markup for each towel? ______

6. Amy Jensen is the chairperson of a charity fund-raising activity. A national direct-sales fund-raising company supplied her with this order form for stationery items. Leslie orders the items at the fund-raiser price. The charity sells the items at a markup of 80% of cost. Overhead is 7% of selling price.

a. Find the selling price for each item. **b.** Find the net profit for each item.

ORDER CODE	NUMBER OF ITEMS	PAGE	CATALOG PRICE	QUANTITY PRICE	FUND-RAISER PRICE	TOTAL AMOUNT	Selling Price	Net Profit
80	20	36	$1.54 ea.	$1.18 ea.	$0.91 ea.		______	______
81	16	9	$3.13 ea.	$2.41 ea.	$1.85 ea.		______	______
82	50	48	$2.47 ea.	$1.90 ea.	$1.46 ea.		______	______
83	26	31	$1.45 ea.	$1.12 ea.	$0.86 ea.		______	______
84	31	19	$2.13 ea.	$1.64 ea.	$1.26 ea.		______	______

Copyright © by Glencoe Division

Name ______________________ Date ______________

LESSON 15-8

Markdown

When you reduce the selling price of an item, the reduction is called a markdown. The markdown rate is expressed as a percent of the regular selling price of the item.

MARKDOWN = REGULAR SELLING PRICE − SALE PRICE

$$\text{MARKDOWN RATE} = \frac{\text{MARKDOWN}}{\text{REGULAR SELLING PRICE}}$$

---

1. Jasper Department Store carried this ad. What is the percent markdown?

Save 5.00
11.99 Reg. 16.99
Men's, Boys' & Youths' Nylon Joggers

____

2. Bright Idea Lamp Shop has lamps on sale as shown. Determine the markdown and percent markdown for each item.

Large selection of lamps
74.99 Bisque table lamp **50.00**
69.99 Pharmacy floor lamp........ **41.99**
49.99 Downbridge floor lamp **29.99**

__________ __________ __________

__________ __________ __________

3. World Travel Luggage Store put this ad in the paper. Compute the percent markdown for each item.

rugged luggage
Was 35.00, tote **24.50**
Was 75.00, 21-in. carry-on **56.25**
Was 95.00, 24-in. pullman **71.25**
Was 105.00, 26-in. pullman **84.00**
Was 140.00, garment bag **119.00**

__________ __________ __________

__________ __________

4. The Sunday newspaper carried an advertisement for All Oak furniture indicating a savings of $65 on a bookcase. The sale price was $314.99. What was the percent markdown? __________

5. Dottie Hanlen works for Delta Grocery Store. She usually works in stock, but because of a shortage of help she has been asked to price some packages of frozen foods. This pricing guide is the only information she has.

Michelle has been asked to mark down the regular selling price of each item 25% for a storewide promotion. What is the regular selling price and the sale price of each package?

DESCRIPTION	CASE COST	PACKAGES PER CASE	MARKUP RATE BASED ON SELLING PRICE	Regular Selling Price	Sale Price
Butter top dinner rolls	$ 9.48	12	25%		
Crescent dinner rolls	7.44	12	29%		
Banana bread	11.76	12	30%		
Apricot nut bread	13.41	9	29%		
Oatmeal nut bread	10.08	12	28%		
Cranberry nut muffins	10.56	8	30%		

Copyright © by Glencoe Division

Name ______________________________ Date ______________

LESSON 16-1

Opinion Surveys

An opinion survey can help you determine how well a product is received by the buying public.

$$\text{RATE OF PARTICULAR RESPONSE} = \frac{\text{NUMBER OF TIMES PARTICULAR RESPONSE OCCURS}}{\text{TOTAL NUMBER OF RESPONSES}}$$

1. The registrar's office conducted a survey on the Student Orientation and Registration (SOAR) program. The 380 students surveyed were asked to choose one answer for this question: "If you consult a counselor, why?" The choices and number of responses received for each were:

95 ______ **a.** Counselor understands situation. ______

146 ______ **b.** Counselor helps me with registration. ______

82 ______ **c.** Counselor gives good advice. ______

57 ______ **d.** Counselor makes referrals for special situations. ______

What is the rate of each response to the nearest tenth of a percent?

2. Your company has decided to conduct an opinion survey to find out how well the new Kitten cat food is selling. A total of 1000 consumers were asked if they would continue to buy the product. They gave the following responses:

	Age Group			
Response	**Under 20**	**20-30**	**30-40**	**Over 40**
Definitely	78	44	84	117
Probably	64	40	61	94
Possibly	51	28	74	106
No	28	18	38	75

a. What is the rate of "No" responses for consumers between 20–30? ______

b. As judged by the rate of "Definitely" responses, which is your most loyal group of customers? ______

c. If a 70% probable or definite response is needed for success, what do you advise with respect to Kitten cat food? ______

3. Scott Park's Auto Service conducted a mail survey of all their clients. In response to the question, "If you do *not* bring your car back to our garage for service, why?", the responses were:

______ 22 Moved away from vicinity

______ 17 Disliked quality of service

______ 34 Service charges too high

______ 39 Crowded service area

______ 63 Location not convenient

______ 15 Some other reason

What is the percent of each response?

Copyright © by Glencoe Division

Name ______________________________ Date ______________

LESSONS 16-2, 16-3

Sales Potential and Market Share

The sales potential of a product is determined by the percent of potential purchasers, the market size, and the individual rate of purchase. The market share is the ratio of the total product sales to the total market sales.

$$\text{ANNUAL SALES POTENTIAL} = \text{PERCENT OF POTENTIAL PURCHASERS} \times \text{ESTIMATED MARKET SIZE} \times \text{INDIVIDUAL RATE OF PURCHASE}$$

$$\text{MARKET SHARE} = \frac{\text{TOTAL PRODUCT SALES}}{\text{TOTAL MARKET SALES}}$$

Round to the nearest tenth of a percent.

	1.	2.	3.	4.
New Product	**Deodorant**	**Computer game**	**Fertilizer**	**Motor oil**
Number in Sample	2000	5000	16,000	8400
Number of Potential Purchasers	725	270	782	265
Percent of Potential Purchasers				
Estimated Market Size	20,000,000	12,000,000	62,000,000	120,000
Individual Rate of Purchase per Year	12 sticks	2 games	4 bags	6 quarts
Annual Sales Potential				

5. Modern Optical Co. is marketing a new style of soft contact lens. Out of a sample of 7000 users, 342 preferred the new style. There is an estimated total market of 875,000 users of contact lenses in the city. The average consumer purchases one pair per year. What is the sales potential for the new lenses for one year? ______________

6. Electronic Air is a new room air freshener. Out of 7500 people surveyed, 82 said they would buy it. The estimated market size is 4,500,000. The company estimates that each person would buy four per year. What is the annual sales potential? ______________

7. Enviro, Incorporated sells approximately 3,516,000 beetle traps per year. The insect control industry sells approximately 9,000,000 beetle traps per year. What is Enviro, Inc.'s market share? ______________

8. Your company sells pen and pencil sets. Last year sales totaled $74.3 million. The total market sales were $3.6 billion. What was your company's market share? ______________

Copyright © by Glencoe Division

Name ______________________________ Date ______________

LESSONS 16-4, 16-5

Sales Projections — Graphs and Factor Method

Sales projections give you an estimate of the dollar volume or unit sales that might occur in the future. You can either use a graph or use the factor method to project sales.

PROJECTED SALES = MARKET-SHARE FACTOR × PROJECTED MARKET SALES

1. Season's Stores sells storm windows. Their sales history shows:

Year	1985	1990	1995	2000
Sales (in millions)	\$2.4	\$2.1	\$3.2	\$3.9

Construct a line graph to project sales for 2005 and 2010.

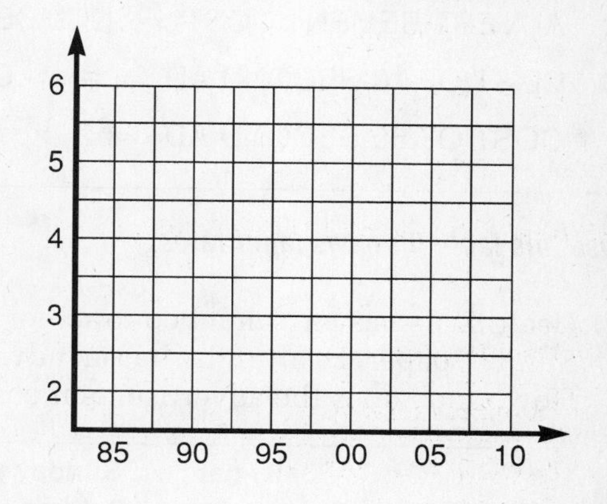

2. At present, Sky Line Foods supplies 46% of the meals for ATB flights, earning an income of \$6,375,000 per year. The estimated ATB market total for next year is \$18,043,533. What is Sky Line Foods' sales projection for next year? ____________

3. You are the Registrar at City College. Your enrollment history for evening courses shows:

	Academic Year			
Semester	1991–92	1992–93	1993–94	1994–95
Fall	1350	1390	1420	1450
Spring	1400	1420	1470	1510

Construct a line graph to project enrollment for both the fall and spring semesters for 1995–96 and 1996–97.

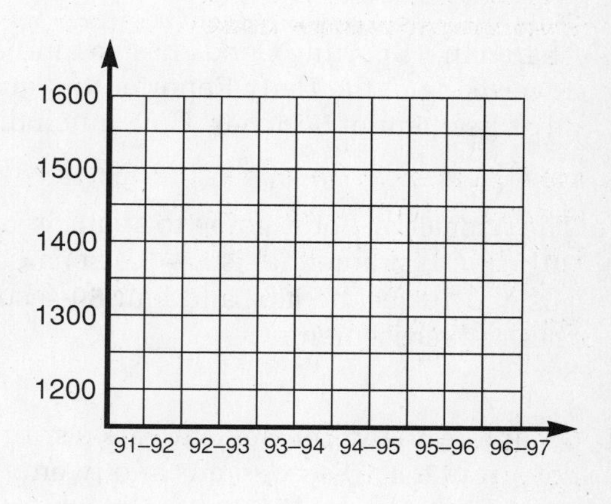

4. Uptown Cleaners has traditionally had 21% of the cleaning business in the village of Westerville. The total estimated cleaning business in Westerville for this year is \$3,060,000. Next year Westerville anticipates a 10% increase in the cleaning business.

a. What business can Uptown Cleaning estimate for this year? ____________

b. What business can Uptown Cleaning project for next year? ____________

Copyright © by Glencoe Division

Name ______________________________ Date ______________

LESSONS 16-6, 16-7

Newspaper and Television Advertising Costs

When you place a newspaper advertisement for your product or service, the cost is determined by the space it occupies and the rate charged per line. When you advertise your product or service on television, the cost depends on the time of day, the program ratings, and the length of the commercial. Television commercials are generally 10, 30, or 60 seconds long.

ADVERTISEMENT COST = NUMBER OF COLUMN INCHES × RATE PER COLUMN INCH

COST OF 10-SECOND AD = $\frac{1}{2}$ × COST OF 30-SECOND AD

COST OF 60-SECOND AD = 2 × COST OF 30-SECOND AD

Use this table for newspaper rates.

1. HealthLine has an annual contract for 126 inches of advertising in the Sunday paper. HealthLine has an advertisement equivalent to 16 inches. How much does the advertisement cost? ______________

Contract	Daily per Column inch	Sunday per Column inch
No contract	$45.54	$55.28
16 inches	$34.90	$42.80
126 inches	$33.90	$41.72

2. Chapman's Sporting Goods has an annual contract for 16 inches of advertising in the Daily Reporter. In Saturday's paper they had an advertisement of 5 inches. How much did the advertisement cost? ______________

3. The Arena Car Lot is going to sponsor arena sporting events. The rate per 30-second commercial is $540. Arena Car Lot contracts for twenty 30-second ads, ten 10-second ads, and four 60-second ads. What is the total cost for these advertisements? ______________

4. Mountain View Amusement Park's spring television advertising campaign will consist of ten 10-second ads and twenty 30-second ads on daytime TV and five 30-second ads and ten 60-second ads on prime-time TV. The rates are $6000 per 30-second daytime and $25,000 per 30-second prime-time ad. What is the total cost of Mountain View 's television campaign? ______________

5. You are in charge of publicity for the town fair next month. You are debating whether to have an advertisement equivalent to 31 column inches in the evening paper or a 26 column inch advertisement in the Sunday paper, or two 30-second TV ads. A 30-second TV ad costs $550. If money is a problem, which one would you select? ______________

What is the cost difference? ______________________________

Copyright © by Glencoe Division

Name ______________________ Date ______________

LESSON 16-8

Pricing

The net income of your sales should be high enough to cover all your expenses and still allow you to make a profit.

$$\text{POSSIBLE NET PROFIT} = \left(\text{SELLING PRICE PER UNIT} - \text{TOTAL COST PER UNIT}\right) \times \text{ESTIMATED UNIT SALES}$$

Round the answers to the nearest cent.

1. Complete the table for Bloomers, a local florist. Which selling price yields the greatest possible profit? ______

How many units of the arrangement should be produced? ______

Selling Price per Unit	Estimated Unit Sales	Total Fixed Costs	Fixed Costs per Unit	Variable Costs per Unit	Total Cost per Unit	Possible Net Profit
$39.99	7000	$120,000		$5.25		
$49.99	5000	$120,000		$5.25		
$44.99	6500	$120,000		$5.25		

2. You are the production manager for Marlin Company. You have assembled these figures about your new Allbright Flashlight. Which selling price yields the greatest possible profit? ______

How many units of the flashlight should be produced? ______

Selling Price per Unit	Estimated Unit Sales	Total Fixed Costs	Fixed Costs per Unit	Variable Costs per Unit	Total Cost per Unit	Possible Net Profit
$ 8.99	15,000	$45,000		$1.35		
$ 9.99	20,000	$45,000		$1.35		
$10.99	25,000	$45,000		$1.35		

3. CompAdd produces microcircuitry boards. They have a fixed overhead of $240,000. The variable cost to produce each board is $0.15 CompAdd assumes they could sell 1,000,000 boards at $0.55 each; 1,500,000 boards at $0.48 each; and 2,000,000 boards at $0.35 each. What selling price will maximize Beta's profits? ______

Selling Price per Unit	Estimated Unit Sales	Total Fixed Costs	Fixed Costs per Unit	Variable Costs per Unit	Total Cost per Unit	Possible Net Profit
$0.55						
$0.48						
$0.35						

Copyright © by Glencoe Division

Name ______________________ Date ______________

Major Foods Corporation: A Simulation

Major Foods Corporation is a major breakfast cereal distributor. Before Major Foods markets a new product, thorough research is conducted to determine the product's market potential. As the vice president for marketing, your job is to incorporate all the research data.

Three months ago Major Foods sent 8000 sample size packages of the new oat cereal CIRCLES to randomly selected consumers in the Northeastern region. Enclosed with the sample was a brief questionnaire. Here are the tabulated responses.

	Age Group			
Response	**Under 18 Years**	**18 to 40 Years**	**40 or Over**	**Total**
Excellent	170	2480	1700	4350
Good	70	377	675	1122
Fair	25	108	449	582
Dislike	5	45	72	122
Total	270	3010	2896	6176

Round to the nearest tenth of a percent.

Response	Total Number of Responses	Rate of Response	Rate of Response: Under 18 Years	18 to 40 Years	40 or Over
Excellent					
Good					
Fair					
Dislike					

Of the 8000 samples mailed, what percent responded?	
What is the overall rate of "excellent" responses for all age groups?	
Of the three different age groups, which one shows the highest percent of "excellent" responses?	
What percent of this age group rated the cereal as "excellent"?	
How many responses were received from this age group in total?	
This age group is what percent of all the responses received?	

Consumers who rate CIRCLES as "excellent" are most likely to purchase the product. Use the age group with the highest percent of "excellent" responses as a guide to project your estimated annual sales potential.

Total Responses of Age Group	Estimated Number of Purchases	Estimated Percent of Purchasers	Estimated Market	Estimated Individual Rate of Purchase/Year	Estimated Annual Sale Potential
			100,000	20 boxes	

Copyright © by Glencoe Division

Name ______________________________ Date ______________

Major Foods Corporation: A Simulation

(CONTINUED)

Assume that the estimated percent of purchasers and the individual rate of purchase per year are national trends. Calculate the estimated national sales potential.

Region	Estimated Market Size	Estimated Annual Sales
Northeast	100,000	
Southern	90,000	
Midwest	110,000	
Pacific Coast	120,000	
Northwest	45,000	
Southwest	80,000	
Alaska	20,000	
Estimated National Annual Sales		

The cost of manufacturing one box of CIRCLES is estimated to be $1.12, which includes direct materials and direct labor. Your overhead is 45% of the selling price. Complete the table to calculate the estimated net profit per box.

Estimated Manufacturing Cost/Box	Overhead Rate	Overhead	Markup	Selling Price	Estimated Net Profit/Box
$1.12	45%		$1.87		

Complete the table to project the estimated national net profit and net-profit rate for CIRCLES.

Estimated Net Profit/Box	Estimated National Annual Sales	Estimated National Net Profit	Estimated National Net-Profit Rate

FINAL THOUGHT

Name some media Major Foods can use to promote the new product.

____

Copyright © by Glencoe Division

Name ______________________ Date ____________

LESSON 17-1

Storage Space

The area occupied by your products or materials until you are ready to use them is known as the storage space. The size of the storage space depends on the size of the item stored.

STORAGE SPACE = VOLUME PER ITEM × NUMBER OF ITEMS

	Item	Carton Dimensions				Number of Items	Storage Space
		Length	Width	Height	Volume		
1.	Cooler	$22\frac{1}{2}$ in.	16 in.	12 in.		35	
2.	Jelly jars (12)	10 in.	12 in.	$5\frac{3}{4}$ in.		200	
3.	Office desk	4 ft	2.5 ft	3 ft		30	
4.	Automobile	11 ft	$6\frac{1}{2}$ ft	$7\frac{3}{8}$ ft		12	

5. A popcorn popper is packed in a carton with dimensions of $13\frac{1}{2}$ in. by $13\frac{1}{2}$ in. by 12 in. How much storage space is required for 40 cartons? ____________

6. A floppy disk storage file is packaged in a box measuring 14 in. by 9 in. by 8 in. Computer Stores, Inc., received a shipment of 12 files. How much storage space is needed? ____________

7. Country Kitchen canning jars are packed 12 to a carton. The carton measures $12\frac{1}{2}$ in. by $9\frac{1}{2}$ by $4\frac{1}{4}$ in. Walton Food Market has placed an order for 1440 jars. How much storage space is required? ____________

Will a warehouse bin measuring 12 ft by 8 ft by 7 ft hold the jars? ____________

8. Teletronics, Inc., manufactures cellular telephone equipment. Each mobile telephone is packaged in a box that is 20 cm by 15 cm by 12 cm. The boxed telephones are then packed 12 to a carton, arranged as shown. Each carton is made of cardboard that is 0.5 cm thick. There are no spacers in the carton. How many cubic meters of space does Teletronics need to store 120 cellular telephones?

____

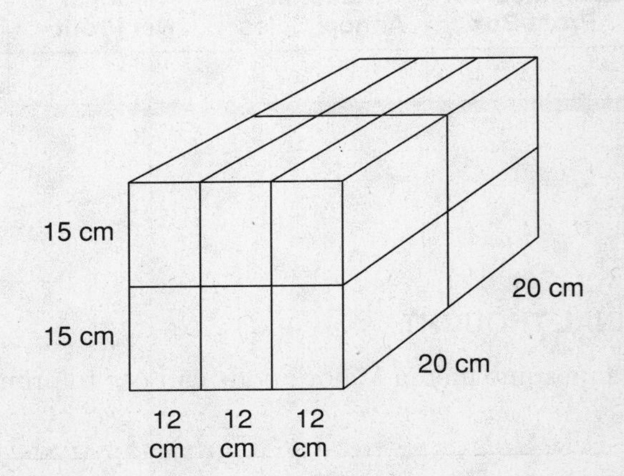

Copyright © by Glencoe Division

Name ______________________________ Date ______________

LESSONS 17-2, 17-3

Taking an Inventory and Valuing an Inventory

To keep a record of all incoming receipts and outgoing items, you can use an inventory to show what is in stock. You can use the average-cost method to calculate the value of your inventory.

INVENTORY = PREVIOUS INVENTORY + RECEIPTS − ISSUES

INVENTORY VALUE = AVERAGE COST PER UNIT × NUMBER ON HAND

1. On July 2 Coulder Rugs had 320 rugs measuring 10 ft by 12 ft in their warehouse. By the end of the month they had shipped out 102 rugs and received 35. How many rugs were on hand on August 1? ____________

2. Lowland Drugstore had 125 boxes of Axin ointment. They issued 20 boxes through sales and sent 52 to a branch store. They did not receive any shipment that week. How many boxes of the ointment remain in stock? ____________

3. Furniture Company orders 50 more model ST-403 end tables. Complete the inventory card below:

Date	Model	Issued	Received	In Stock
9/15	ST-403	6	10	5
9/28	ST-403	—	50	

4. Records for Ingman's Deli show this opening balance and these receipts for Ingman's Mustard from January through July. At the end of July 52 jars were on hand.

What is the value of the inventory? ____________

Date	Receipts	Unit Cost	Total Cost
3/1	60	$1.42	$85.20
4/4	20	1.60	32.00
5/1	40	1.55	62.00
6/2	50	1.45	72.50
7/1	55	1.40	77.00
Total			

5. You work for Bartlett's Maintenance Supplies. Every three months you have to update the inventory and report the value of the inventory on hand. What is the value of the inventory for the fiber glass tubs if only 15 are left? ____________

Date	Receipts	Unit Cost	Total Cost
6/1	35	$196.80	______
7/1	40	204.90	______
8/1	60	209.10	______
Total	______	______	______

Copyright © by Glencoe Division

Name ______________________ Date ______________

LESSON 17-4

Carrying an Inventory

The cost of keeping a sufficient inventory of goods includes taxes, insurance, storage fees, and handling charges. The annual cost of carrying an inventory is often expressed as a percent of the inventory value.

ANNUAL COST OF CARRYING INVENTORY = PERCENT × INVENTORY VALUE

1. New Vision Camera Shop maintains a $120,000 inventory. The cost of keeping the merchandise in stock is 22% of the value of the inventory. What is the annual cost of carrying the inventory? ______

2. Lone Pine Furniture maintains a $246,250 inventory. The cost of keeping a sufficient inventory is 32% of the value of the inventory. What is the annual cost of carrying the inventory? ______

3. OPER Laboratories estimates the cost of carrying its inventory of chemicals and apparatus to be 40% of the value of the merchandise. How much does it cost OPER annually to carry a $124,000 inventory? ______

4. Warehouse Food Market estimates the annual cost of carrying its inventory at 25% of the value of the inventory as shown. What is the annual cost of each expense if the inventory is valued at $410,000?

Type of Expense	Percent	Annual Cost
Spoilage and physical deterioration	7.0%	______
Interest	5.9%	______
Handling	3.0%	______
Storage facilities	1.8%	______
Transportation	1.2%	______
Taxes	0.6%	______
Insurance	5.5%	______
Total	25%	______

5. The rate of inventory turnover is the number of times during one year that a business sells its average inventory. Generally, the higher the rate of inventory turnover, the better the business is doing.

$$\text{INVENTORY TURNOVER} = \frac{\text{TOTAL COST OF GOODS SOLD}}{\text{AVERAGE VALUE OF INVENTORY}}$$

Calculate the inventory turnover for each business.

	a.	b.	c.
Name	Liberty Fabrics	Loren Industrial	General Stores, Inc.
Total Cost of Goods Sold During Year	$141,060	$410,295	$930,000
Average Value of Inventory During Year	$ 31,040	$156,000	$310,000
Inventory Turnover			

Copyright © by Glencoe Division

Name ______________________________ Date ______________

LESSONS 17-5, 17-6

Transportation by Air and Truck

Your business may ship merchandise to customers by air freight or by truck. The shipping charges are paid either by your business or by the customer.

TOTAL SHIPPING COST BY AIR FREIGHT = PICKUP CHARGE + AIR FREIGHT CHARGE + DELIVERY CHARGE + FEDERAL TAX

TOTAL SHIPPING COST BY TRUCK = WEIGHT × BASIC RATE

Round to the nearest cent.

1. A crate of figurines that weighs 240 pounds is shipped by air freight. The total cost of pickup and delivery is $50. The air freight rate is $67.50 per 100 pounds, charged on the actual number of pounds shipped. A 5% federal tax on the air freight charge is applied. What is the total shipping cost? ____________

2. A 430-pound shipment of brass lamps is shipped by truck. The distance from the factory to the store is 356 miles. Use the table on page 462 of your textbook to find the total shipping cost. ____________

3. Norberg's Greenhouse is shipping tulip bulbs to a customer by air freight. The pickup charge is $7.50, and the delivery charge is $8.25. The air freight rate is $71.50 per 100 pounds, charged on the actual number of pounds shipped. A 5% federal tax is applied, based on the freight charge. What is the total cost if the shipment weighed 125 pounds? ____________

4. Eve Meyers is shipping 2 trunks to her college by truck. The distance from her home to the college is 424 miles. The trunks weigh a total of 130 pounds. Use the table on page 462 of your textbook to find the total shipping cost. ____________

5. You have to ship a package 150 miles to a customer. The air freight charge is $58.85 per 100 pounds, charged on the actual number of pounds shipped. The total pickup and delivery charge is $40. No federal tax is applied. How much is the shipping cost if the package weighs 45 pounds? ____________

How much would it cost to send the package by truck? ____________

6. Many businesses ship small items through the mail. The cost of shipping a package by parcel post within the United States depends on its weight and the zones to which and from which it is being shipped. The cost of shipping a package to another country is uniform throughout the United States and is based on the weight and destination of the package.

Designer's Workshop, a mail-order house, is sending several packages by parcel post to Frankfort, Germany. The parcel post rate is $5.60 for the first 2 pounds and $0.95 for each additional pound or fraction of a pound. Find the total cost of shipping each item.

	Weight	Cost for First 2 Pounds	Additional Weight	Cost for Additional Weight	Total Cost
a.	$16\frac{1}{4}$ lb	$5.60	$14\frac{1}{4}$	15 × $0.95 = $14.25	$19.85
b.	$9\frac{1}{2}$ lb				
c.	$1\frac{3}{4}$ lb				

Copyright © by Glencoe Division

Name ______________________________ Date ______________

LESSON 18-1

Building Rental

Your business may rent or lease space in a building on an annual basis. The rent is based on a certain rate per square foot per year. Your total monthly rental charge depends on the number of square feet that your business occupies.

$$\text{MONTHLY RENTAL CHARGE} = \frac{\text{ANNUAL RATE} \times \text{NUMBER OF SQUARE FEET}}{12}$$

Find the monthly rental charge.

	Type of Building	Dimensions	Number of Square Feet	Annual Rate per Square Foot	Monthly Rental Charge
1.	Pharmacy	25 ft by 40 ft		$10.50	
2.	Dry Cleaners	60 ft by 30 ft		$ 6.75	
3.	Gift Shop	35 ft by 25 ft		$ 8.25	

4. The Bargain Barn is considering the rental of additional space at $5.85 per square foot per year. The space Bargain Barn wants to rent measures 150 feet by 150 feet. What monthly rent will Bargain Barn pay for the additional space? ______________

5. Brookmans has rented additional mall space to expand its toy operation. The space measures 20 feet by 40 feet and rents for $15.75 per square foot per year. What monthly rent does Brookmans pay for the additional space? ______________

6. Tops Sport Shop is opening a store at the Eastland Mall. The rent is $18.95 per square foot per year plus $5\frac{1}{2}$% of the store's gross sales. The area of the store is 2500 square feet. If Tops has $526,000 in gross sales the first year, what monthly rent will it pay? ______________

7. Sutter's Fireplace is opening a branch store. The area of the new store is 4100 square feet. The annual rental charge is $9.65 per square foot per year, plus $4\frac{1}{2}$% of Sutter's gross sales for the first year. If Sutter's has $510,000 in gross sales the first year, what monthly rent will it pay? ______________

8. Your company is looking for a new warehouse to store the incoming inventory. At the Hurt Building, you can rent a space measuring 2500 square feet for $9.75 per square foot per year. At the Maine Plaza Warehouse, you can rent a space measuring 3000 square feet for $9.25 per square foot per year. How much is the monthly rental charge at the Hurt Building? ______________

How much is the monthly rental charge at the Maine Plaza Warehouse? ______________

Which location has a lower rental charge? ______________

Copyright © by Glencoe Division

Name ______________________________ Date ______________

LESSON 18-2

Maintenance and Improvement

The total cost of keeping your building clean and maintained generally includes a labor charge and a materials charge. The labor charge is usually calculated on an hourly basis for each service person involved.

TOTAL CHARGE = LABOR CHARGE + MATERIALS CHARGE

1. Liverpool Real Estate is moving to another location in the city. It takes 4 people 7 hours to complete the move. The hourly rate per person is $9.25 in addition to a packing charge of $54.50. What is the total moving charge? ______________

2. Piazza Pizza Shop hired 3 carpenters to remodel its store. The carpenters each earn $21.90 per hour. Each carpenter worked 161_2 hours. The materials charge was $3160.40. What was the total charge? ______________

3. Long Haul Moving Company needs to reseal their asphalt driveway. It will take 2 workers 6 hours to complete the job, and 50 gallons of tar sealer. The charge for the tar sealer is $10.59 per 5-gallon can. The labor cost is $11.25 per person per hour. What is the total cost? ______________

4. The management of Lake View Condominiums hired 2 people to repaint the main lobby. It took the painters 7 hours to finish the job. They used 4 gallons of paint at $17.99 per gallon. The painters charged $14.50 per hour per person for labor. How much did the management pay to repaint the lobby? ______________

5. Your company had minor flood damage. You received these estimates for repair work from Faber Repair.

	Hourly Rate	Time Required	Materials Charge
Electrician	$24.10	5 hr	$112.90
Carpenter	$21.60	$8\frac{1}{2}$ hr	$268.40
Plumber	$19.25	10 hr	$ 54.60

You must decide if you should hire these 3 people or have your own employees do the job. Materials would cost you $450.10. It would take 4 of your employees 8 hours each to complete the job. Three of your employees earn $8.95 per hour, and one earns $10.95 per hour. If the price is the only consideration, which would be less expensive? ______________

6. Shelby Cleaning Service, Inc., is retained to clean the Hunton Law Clinic daily from Monday through Saturday. Two Shelby Cleaning employees work at the clinic for 2 hours each day. One employee is paid $5.25 per hour while the other employee is paid $7.50 per hour. Both earn double time on Saturday. Shelby Cleaning adds 55% of the cost of labor to cover overhead. What does Shelby Cleaning Service charge the Hunton Law Clinic per week? ______________

Copyright © by Glencoe Division

Name ______________________________ Date ______________

LESSON 18-3

Equipment Rental

The cost of renting equipment or furniture for your business is based on the rental charge and the length of time for which you rent them. In some states, a sales tax is added to the rental cost.

TOTAL RENTAL COST = (RENTAL CHARGE × TIME) + TAX

	Item	List Price	Monthly Charge	Monthly Charge	Time	Tax	Rental Cost
1.	Stereo	$1490	10% of list		12 months	6%	
2.	Color TV	$1010	6% of list		9 months	$5\frac{1}{2}$%	
3.	Computer Equipment	$5940	10% of list		3 months	8%	
4.	Wheelchair	$ 495	12% of list		6 months	6.5%	

5. Wendell Construction plans to rent a 300 horsepower bulldozer for 3 months. The rental charge is $41,872.50 per month, without an operator. The delivery and pickup charge amounts to $290 (no tax on delivery and pickup). If the list price for this bulldozer is $465,250, what percent is the monthly rental charge? ______

What is the total rental charge at the end of 3 months, if a $6\frac{1}{2}$% tax is applied? ______

6. The Hunter and Williams law office is renting additional furniture for the next 8 months. The rental charge is 10% of the list price of the new furniture per month. There is a $6\frac{1}{2}$% state tax. They plan to rent:

Item	List Price
4 leather arm chairs	$ 510 each
1 oak conference table to seat 15 people	$9680 each
1 Fax machine	$ 350 each
1 copy machine	$5080 each

What is the total rental charge? ______

7. Your office needs to rent 3 office trailers for 4 months. You received cost information from 2 rental agencies. If price is the only factor, which company should you contact? ______

Dena's Rental	Neuman Supply
Per month:	Per month:
$500 per trailer	$450 per trailer
5% insurance charge	no insurance
5.5% use tax	6% use tax
no transportation charge	$25 delivery charge per office trailer

Copyright © by Glencoe Division

Name ______________________________ Date ______________

LESSONS 18-4, 18-5

Utilities Costs: Telephone and Electricity

Your monthly telephone charge depends on the number of incoming lines, the type of equipment, and the type of service your business uses. A federal excise tax is added to your telephone charge each month. The monthly cost of electricity depends on the demand charge and the energy charge. The fuel adjustment charge may be added by the electric company to help cover the fluctuating cost of generating electricity.

TOTAL TELEPHONE COST FOR MONTH = BASIC MONTHLY CHARGE + COST OF ADDITIONAL CALLS + FEDERAL EXCISE TAX

TOTAL ELECTRICITY COST FOR MONTH = DEMAND CHARGE + ENERGY CHARGE + FUEL ADJUSTMENT CHARGE

1. Seaway Corporation has 5 telephones for a total basic monthly charge of $112.90. The charge includes 400 local outgoing calls. The cost of additional local calls is $0.09 each. Last month, 619 outgoing calls were made. In addition, long distance calls totaling $210.90 were made. A 5% federal excise tax is charged on the basic charge, extra local calls, and long distance calls. What is the total cost for the month? ______________

2. Continental Bank's suburban branch has 2 incoming lines. Each has a digital touch-tone telephone. The branch also has 5 extension telephones. The monthly charge includes 275 outgoing local calls. Each additional local call costs $0.13. The branch made 390 outgoing local calls last month. The federal excise tax is 5%. What is the cost of telephone service? ______________

Monthly Charge Per Phone	
Digital Telephone	$18.90
Extension Telephone	5.10
Touch-Tone Service	4.05

Refer to the chart for Problems 3–5.

NORTHWEST POWER	
Demand Charge	
First 50 kW	$3.95/kW
Over 50 kW	3.80/kW
Energy Charge	
First 250 kW•h	$0.085/kW•h
Next 750 kW•h	0.069/kW•h
Next 2000 kW•h	0.05/kW•h
Next 2000 kW•h	0.039/kW•h
Next 5000 kW•h	0.035/kW•h
Over 10,000 kW•h	0.04/kW•h
Fuel adjustment	$0.0215/kW•h

3. Kilgan's Grocery used 13,000 kW•h of electricity last month. The peak load for the month was 80 kW. What is the total cost of electricity for Kilgan's Grocery? ______________

4. P&J Printing used a total of 25,480 kW•h of electricity for the month. The peak load during the month was 135 kW. What is the total cost of electricity for the month? ______________

5. You own the Sunrise Cafe. Last month you used 9700 kW•h of electricity with a peak load of 90 kW. In addition to your monthly electricity bill, you have your telephone bill to pay. You have 2 telephones for a total basic monthly charge of $42.50. The charge includes 180 local outgoing calls. The cost of additional calls is $0.08 each. Last month, 215 outgoing calls were made. Long distance calls totaled $149.10. A 5% federal excise tax is charged.

a. Calculate last month's telephone cost to the nearest cent. ______________

b. What was last month's electricity bill? ______________

Copyright © by Glencoe Division

Name ______________________ Date ______________

LESSON 18-6

Professional Services

Your business may seek professional advice on a particular problem. Some consultants charge a flat fee; some charge a percent of the cost of the project; and some charge by the hour.

TOTAL COST = SUM OF CONSULTANTS' FEES

	Professional Service	Fee Structure	Project Information	Total Fee
1.	Architect	8% of project cost	$925,000	
2.	Real Estate Developer	5.5% of project cost	$510,000	
3.	Contractor	$19.25 per hour	25 hours worked	
4.	Engineer	$54.00 per hour	25 hours worked	

5. Riverview High School hired an in-service training specialist to conduct workshops for all their science teachers. The specialist charged a flat fee of $300 per day. The workshops lasted $2\frac{1}{2}$ days. What was the total cost of the specialist's services? ______

6. Conner and Helwig, an insurance firm, plans to issue $58,000,000 in bonds to pay for an extensive expansion. One bond broker will sell the bonds for a fee of 9.25% of the $58,000,000 face value of the bonds. What will the broker's services cost the firm? ______

7. Maxwell's, a discount store, wants to change their store's image. They hire Century Interiors to redecorate their store. Blanchard charges $3500 plus $16.25 per hour for each of the 3 designers assigned to the project. It takes Blanchard designers 384 hours each to complete the design. How much does this project cost Maxwell's in total? ______

8. Your company has purchased prime real estate in the financial district. You hired an attorney to handle the legal transactions. You were charged $115 per hour plus a $6500 fee. The attorney devoted 490 hours to the project. How much did your company pay for these legal services? ______

9. General Hospital hired an industrial engineering firm to conduct a work sampling of the average nurse's day. I.E., Inc., did the work sampling and charged $86.50 per hour. It took I.E. 36 hours to complete the task. What did the work sampling cost General Hospital? ______

10. Save Rite has installed laser readers at all their checkout stations in all of their 5 stores. Save Rite hired Megan Alexander to consult on the equipment and its installation. Alexander charged 9.5% of the total cost. Save Rite hired Don Engel to instruct its employees on the use of the equipment. Engel conducted a 4-hour session at each store. Engel charges $68 per hour. The cost of the equipment and installation totaled $519,480. What was the total cost for professional services? ______

Copyright © by Glencoe Division

Name ____________________ Date ____________

LESSON 19-1

Payroll

A payroll register is a record of the gross income, deductions, and net income of your company's employees.

Use the tables on pages 642–643 of your textbook for federal withholding tax (FIT). Use the social security tax rate of 6.2% and medicare tax rate of 1.45%.

1. Donna's Cookie Factory employs 4 people to make deliveries. Donna's pays an hourly rate of $5.50. The only deductions are federal withholding, social security, and medicare. Complete the payroll register for the week.

Employee	Income Tax Information	Hours Worked	Gross Pay	FIT	Soc. Sec.	Med.	Total Ded.	Net Pay
Weinberg, A.M.	S, 1 allow.	40						
Smythe, P.	M, 2 allow.	36						
Penzo, C.	S, 0 allow.	32						
Sheen, N.	S, 1 allow.	34						
	Total							

2. Your job is to complete the payroll register for Sport Shoes. A total of 6 salespeople are employed. They are each paid $5.30 per hour. The only deductions are federal withholding, social security of 6.2%, medicare of 1.45%, and a city income tax (CIT) of 0.7% of the income.

Employee	Income Tax Information	Hours Worked	Gross Pay	FIT	Soc. Sec.	Med.	CIT	Total Ded.	Net Pay
Clark	S, 0 allow.	32							
Elliot	M, 3 allow.	40							
Grant	M, 1 allow.	36							
King	S, 2 allow.	35							
Peters	S, 1 allow.	38							
Rogers	M, 2 allow.	32							
	Total								

Copyright © by Glencoe Division

Name ____________________ Date ____________

LESSON 19-2

Business Expenses

Your business must keep accurate records of all its expenses. This information can be used to calculate profit and income tax or to plan future spending. You may calculate the percent of each expense by comparing it to your total expenses.

$$\text{PERCENT OF TOTAL} = \frac{\text{PARTICULAR EXPENSE}}{\text{TOTAL EXPENSES}}$$

1. Holman's Florist and Garden Shop had the following expenses last quarter. Find the total. Find what percent each expense is of the total.

Item	Amount	Percent of Total
Payroll	$126,250	
Advertising	25,500	
Equipment	18,900	
Office/garage rental	6,100	
Supplies/flowers	110,250	
Insurance	8,800	
Utilities	6,300	
Total		

2. Your business had the following expenses for the last quarter. Find the total. Find what percent each expense is of the total.

Item	Amount	Percent of Total
Payroll	$23,000	
Advertising	7,000	
Raw materials	10,480	
Equipment rental	9,960	
Factory rental	9,850	
Office rental	2,100	
Warehouse rental	980	
Office supplies	860	
Utilities	510	
Total		

Copyright © by Glencoe Division

Name ______________________ Date ______________

LESSON 19-3

Apportioning Expenses

Your business may apportion certain expenses among its departments. Often, the amount that each department is charged depends on the space that it occupies.

$$\text{AMOUNT PAID} = \frac{\text{SQUARE FEET OCCUPIED}}{\text{TOTAL SQUARE FOOTAGE}} \times \text{TOTAL EXPENSE}$$

1. The Braden Building Corporation apportions the annual cost of utilities among its departments on the basis of space occupied. The total utilities cost for the year was $564,210. The total area of the building is 680,000 square feet. The accounting department occupies an area that is 50 feet by 65 feet. What amount was the accounting department charged for utilities for the year? ______________

2. The cost of heating Lentel Plastics totaled $58,460 for the winter. Lentel apportions this expense among its departments. If the company occupies a total of 18,900 square feet, how much did the following departments pay for heating?

Mailroom, 35 ft by 26 ft ______________ Accounting, 60 ft by 40 ft ______________

3. Your business apportions cost among the departments on the basis of gross sales. The gross sales for last year totaled $4,650,900. Some of the annual expenses were distributed as follows:

Maintenance	Utilities	Security
$5600	$16,980	$30,600

The Furniture Department had $490,500 in gross sales last year. How much did it pay for each of the annual expenses? ______________

The Home Furnishing Department had $812,000 in gross sales last year. How much did it pay for each of the annual expenses? ______________

4. You are the manager for the XCEL Development Company that operates the Seasons Shoppers Mall. XCEL apportions the cost of the various items based on the square footage occupied by each store. The utilities cost $90,550, the ads cost $17,600, and the maintenance cost is $154,000. You are to apportion each expense.

Store	Dimensions (in feet)	Square Footage	AMOUNT OF EACH EXPENSE		
			Utilities	Advertising	Maintenance
Ice Cream Parlor	15 × 20				
Jem's Jewelers	45 × 65				
Jeans Shoppe	40 × 90				
Shoe Barn	30 × 80				
Alley Gifts	30 × 80				
Toy Town	45 × 65				
Kite Corner	15 × 25				
Total					

Copyright © by Glencoe Division

Name ______________________ Date ______________

Depreciation — Straight-Line Method and Book Value

For tax purposes, the Internal Revenue Service allows you to recognize the depreciation of many of the items that your business owns. The straight-line method is one way to calculate the annual depreciation of an item. The book value is the approximate value of an item after you have owned it for a while.

$$\text{ANNUAL DEPRECIATION} = \frac{\text{ORIGINAL COST} - \text{RESALE VALUE}}{\text{ESTIMATED LIFE}}$$

$$\text{BOOK VALUE} = \text{ORIGINAL COST} - \text{ACCUMULATED DEPRECIATION}$$

1. Use the straight-line method of depreciation to construct depreciation records for computer equipment that cost $9400, has an estimated life of 3 years, and has an estimated resale value of $1000.

End of Year	Calculations for Depreciation	Annual Depreciation	Accumulated Depreciation	Book Value
1				
2				
3				

2. Larry Borgen is a computer programmer. He purchased equipment for his department for $13,990. The resale value of the equipment is estimated to be $725 after 5 years of use. Use the straight-line method to complete the depreciation record.

End of Year	Calculations for Depreciation	Annual Depreciation	Accumulated Depreciation	Book Value
1				
2				
3				
4				
5				

3. Your business is planning to buy a phone mail system. The cost is $15,600. The resale value is estimated to be $1200 after 5 years of use. Use the straight-line method to complete the depreciation record.

End of Year	Calculations for Depreciation	Annual Depreciation	Accumulated Depreciation	Book Value
1				
2				
3				
4				
5				

Copyright © by Glencoe Division

Name ______________________ Date ____________

LESSON 19-6

Sum-of-the-Year's-Digits Method

The sum-of-the-years'-digits method is another way you can determine the annual depreciation of an item. To use this method, you must know the original cost, the estimated life, and the resale value.

$$\text{DEPRECIATION FOR YEAR} = \text{FRACTION OF TOTAL DEPRECIATION} \times \left(\text{ORIGINAL COST} - \text{RESALE VALUE}\right)$$

1. Use the sum-of-the-years'-digits method to construct depreciation records for a minicomputer that cost $25,000, has an estimated life of 3 years, and an estimated resale value of $1000.

End of Year	Calculations for Depreciation	Annual Depreciation	Accumulated Depreciation	Book Value
1				
2				
3				

2. Use the sum-of-the-years'-digits method to construct a depreciation record for a restaurant oven that cost $4800, has an estimated life of 5 years, and an estimated resale value of $300.

End of Year	Calculations for Depreciation	Annual Depreciation	Accumulated Depreciation	Book Value
1				
2				
3				
4				
5				

3. Your company recently purchased a new computerized accounting system for a total cost of $18,975. The system has an estimated life of 5 years. The resale value after 5 years is expected to be $2100. Use the sum-of-the-years'-digits method to construct a depreciation record.

End of Year	Calculations for Depreciation	Annual Depreciation	Accumulated Depreciation	Book Value
1				
2				
3				
4				
5				

Copyright © by Glencoe Division

Name ______________________ Date ______________

LESSON 19-7

Double-Declining-Balance Method

Your business may use the double-declining-balance method to find the annual depreciation of an item. To use this method, you must know the original cost, the estimated life of the item, and the resale value.

$$\text{ANNUAL DEPRECIATION RATE} = 2 \times \frac{100\%}{\text{ESTIMATED LIFE}}$$

$$\text{DEPRECIATION FOR YEAR} = \text{ANNUAL DEPRECIATION RATE} \times \text{PREVIOUS DECLINING BALANCE}$$

1. A new car that is used for business purposes costs $16,800. The car has an estimated life of 4 years and a resale value estimated at $1160. Construct a depreciation record using the double-declining-balance method of depreciation.

End of Year	Calculations for Depreciation	Annual Depreciation	Accumulated Depreciation	Book Value
1				
2				
3				
4				

2. Superline Grocery purchased new checkout equipment. The original cost was $68,000. The life of the equipment is estimated to be 5 years. The resale value is expected to be $5500. Construct a depreciation record using the double-declining-balance method of depreciation.

End of Year	Calculations for Depreciation	Annual Depreciation	Accumulated Depreciation	Book Value
1				
2				
3				
4				
5				

3. Your business purchased new computer accounting software for $10800. The resale value is expected to be $0 after a 4-year life. Construct a depreciation record using the double-declining-balance method of depreciation.

End of Year	Calculations for Depreciation	Annual Depreciation	Accumulated Depreciation	Book Value
1				
2				
3				
4				

Copyright © by Glencoe Division

Name ______________________ Date ______________

LESSON 19-8

Modified Accelerated Cost Recovery System (MACRS)

The Modified Accelerated Cost Recovery System (MACRS) is another method of computing depreciation.

ANNUAL DEPRECIATION = FIXED PERCENT × ORIGINAL COST

BOOK VALUE = ORIGINAL COST − ACCUMULATED DEPRECIATION

1. Use the MACRS method to complete the depreciation record for a $17,900 car used for business purposes. Autos can be depreciated 33% the first year, 45% the second year, 15% the third year, and 7% the fourth year.

End of Year	Calculations for Depreciation	Annual Depreciation	Accumulated Depreciation	Book Value
1				
2				
3				
4				

2. Use the MACRS method to complete the depreciation record for a $90,460 computer. Computers can be depreciated 26% the first year, 32% the second year, 19.20% the third year, 11.52% the fourth year, and 11.52% the fifth year, and 5.76% the sixth year.

End of Year	Calculations for Depreciation	Annual Depreciation	Accumulated Depreciation	Book Value
1				
2				
3				
4				
5				
6				

3. Your business purchased 10 new desk top computers and a networking system for a total cost of $32,000. Use the MACRS method to complete the depreciation record for 5 years. Use the percents in exercise 2.

End of Year	Calculations for Depreciation	Annual Depreciation	Accumulated Depreciation	Book Value
1				
2				
3				
4				
5				
6				

Copyright © by Glencoe Division

Name ______________________ Date ______________

LESSON 20-1

Assets, Liabilities, and Equity

Assets are the total of your cash, the items that you have purchased, and any money that your customers owe you. Liabilities are the total amount of money that you owe the creditors. Owner's equity, net worth, or capital is the total value of the assets that you own outright, after you subtract your liabilities.

OWNER'S EQUITY = ASSETS − LIABILITIES

	1.	2.	3.	4.	5.
Assets	$68,410.04	$94,014.30	$140,000.00	$76,494.12	$9140.00
Liabilities	$10,243.00	$18,480.10	$45,630.40	$21,410.61	$8721.49
Owner's Equity					

6. Dot's Dress Shop has these assets and liabilities:

Cash:	$2160.40	Supplies:	$270.90	Unpaid merchandise:	$5614.20
Inventory:	$10,427.60	Building:	$62,000.00	Taxes owed:	$912.07
Equipment:	$9160.14	Land:	$16,000.00	Real estate loan:	$16,480.90

What are the total assets? ______

What are the total liabilities? ______

What is the owner's equity? ______

7. First Start Day Care has these assets and liabilities:

Cash:	$310.90	Supplies:	$416.80	Taxes owed:	$2160.10
Tuition:	$21,000.00	Fixtures:	$3,487.95	Wages owed:	$3140.90
Rent owed:	$1,600.00	Bank loan:	$12,910.00		

What are the total assets? ______

What are the total liabilities? ______

What is the owner's equity? ______

8. You are the owner and operator of the Southside Dry Cleaners. You have the following assets and liabilities:

Cash:	$2,640.10	Supplies:	$3,100.00	Accounts Payable:	$4,110.90
Inventory:	$576.90	Building:	$67,800.00	Taxes owed:	$2,160.80
Equipment:	$26,040.00	Land:	$21,000.00	Real estate loan:	$59,640.10

What are your total assets? ______

What are your total liabilities? ______

What is your equity? ______

Copyright © by Glencoe Division

Name ______________________ Date ____________

LESSON 20-2

Balance Sheet

A balance sheet shows the financial position of your company on a certain date. You may prepare a balance sheet monthly, quarterly, or annually. The balance sheet shows your total assets, total liabilities, and owner's equity. The sum of the assets must equal the sum of the liabilities and owner's equity.

1. Complete the balance sheet for Morton Auto Parts Inc.

Morton Auto Parts
Balance Sheet
November 30, 19—

Assets		Liabilities	
Cash on hand	4719.10	Bank loan	21090.60
Accts. receivable	78400.90	Accts. payable	15490.10
Inventory	18101.40	Taxes owed	1710.80
Supplies	416.80	Wages owed	612.19
Store fixtures	1816.10	Mortgage loan	64910.00
Building	86410.00		
Land	21000.00	Total Liabilities	
		Owner's Equity	
		Capital	
		Total Liabilities	
Total Assets		and Owner's Equity	

2. You are the owner of the Seashore Gift Store. You had the following assets and liabilities on July 31. Complete the balance sheet for your gift shop.

Assets	
Cash	$2,160.90
Accounts receivable	$1,914.06
Inventory	$37,519.00
Prepayments	$2,160.00
Property	$93,575.00
Investments	$7,500.00
Other assets	$5,617.81

Liabilities	
Accounts payable	$12,070.50
Notes payable	$65,410.10
Income taxes	$2,817.00
Other liabilities	$5,160.90

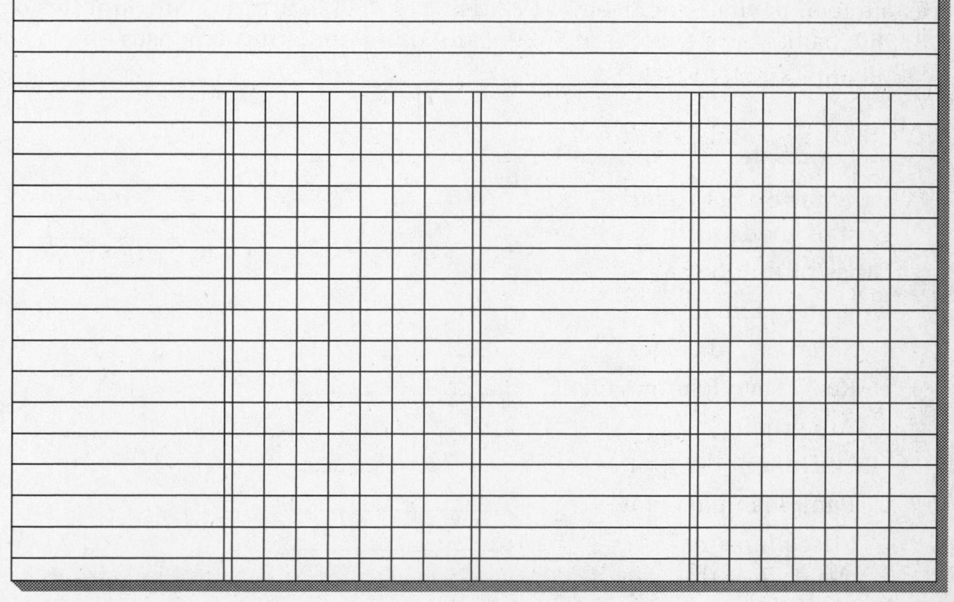

Copyright © by Glencoe Division

Name ______________________ Date ______________

Cost of Goods Sold and Income Statements

An income statement, or profit-and-loss statement, shows in detail your income and operating expenses. If your gross profit is greater than your total operating expenses, your income statement will show a net income, or net profit.

GROSS PROFIT = NET SALES − COST OF GOODS SOLD

NET INCOME = GROSS PROFIT − TOTAL OPERATING EXPENSES

COST OF GOODS SOLD = (BEGINNING INVENTORY + RECEIPTS) − ENDING INVENTORY

	Total Sales	Returns	Net Sales	Cost of Goods Sold	Gross Profit	Operating Expenses	Net Income
1.	$ 61,430.00	$9460.00	$ 51,970.00	$20,788.00	$ 31,182.00	$ 2,618.00	
2.	$ 30,812.90	$3475.00		$ 9,214.00		$ 6,480.90	
3.	$646,900.00	$6890.00		$81,114.00		$28,910.00	

Complete an income statement for each business in Problems 4–5.

4. Last month, Ashton Auto Parts, Inc., had total sales of $16,914.06. Merchandise totaling $312.04 was returned. The auto parts that were sold cost Ashton Auto Parts $8608.00. Operating expenses for the month were $1961.02.

Total Sales	Returns	Net Sales	Cost of Goods Sold	Gross Profit	Operating Expenses	Net Income

5. You prepare a quarterly income statement for Duncan Manufacturing Company. For this past quarter, Duncan Manufacturing Company had total sales of $212,090, and returns of $6187. The cost of goods sold amounted to $94,860. Operating expenses for the quarter included: salaries and wages of $55,410, real estate loan payment of $5440, advertising at $5000, utilities and supplies of $3960, bank loan payment of $3990, and other operating expenses of $4660.

Income:
- Sales ______
- Less: Sales returns ______
- Net sales ______

Cost of goods sold ______

Gross profit on sales ______

Operating Expenses:
- Salaries and wages ______
- Real estate loan payment ______
- Advertising ______
- Utilities and supplies ______
- Bank loan payment ______
- Miscellaneous ______
- Total operating expenses ______

Net Income ______

Copyright © by Glencoe Division

Name ______________________________ Date ______________

LESSON 20-5

Vertical Analysis

Your business may analyze the dollar amount for the items on its income statement by converting each amount to a percent of net sales.

$$\text{PERCENT OF NET SALES} = \frac{\text{AMOUNT FOR ITEM}}{\text{NET SALES}}$$

---

1. Duncan Manufacturing had net sales of $205,903. The cost of goods sold by Duncan was $94,860. What is the gross profit as a percent of net sales? ______________

2. Ashton Auto Parts had total sales of $16,914.06. Returns totaled $312.04. The auto parts sold cost Ashton $8608. What were Sunside's net sales? ______________

What was Ashton's gross profit on sales? ______________

What is the gross profit as a percent of net sales? ______________

3. The income statement for Century Computer Mart for one month showed these figures.

Net sales	$156,910
Cost of goods sold	84,520
Gross profit on sales	

What is the gross profit on sales? ______________

What is the gross profit as a percent of net sales? ______________

4. You are the proprietor of Hi-Fi Sounds. The income statement for your corporation for the past quarter showed these figures.

Net sales	$84,610
Cost of goods sold	41,010
Gross profit on sales	

Total operating expenses	$16,100
Net Income	

What is the gross profit on sales? ______________

What is the cost of goods sold as a percent of net sales? ______________

What is the gross profit on sales as a percent of net sales? ______________

What are the operating expenses as a percent of net sales? ______________

What is the net income? ______________

What is the net income as a percent of net sales? ______________

Copyright © by Glencoe Division

Name ______________________________ Date ____________

LESSON 20-6

Horizontal Analysis

You can compare income statements by computing percent changes from one income statement to another. When you compute the percent change, the dollar amount on the earlier statements is the base figure. The amount of change is the difference between the base figure and the figure on the current statement.

$$\text{PERCENT CHANGE} = \frac{\text{AMOUNT OF CHANGE}}{\text{BASE FIGURE}}$$

Round to the nearest tenth of a percent.

	Last Year (Base)	This Year	Amount of Change	Percent Change
1.	$910,400	$847,000	−$ 63,400	
2.	$252,000	$296,000	+$ 44,000	
3.	$ 99,900	$ 93,418		
4.	$985,593	$798,410		

5. Income statements for Williams, Barnes, and Sampson, Inc., showed these figures for June and July. Complete the table.

	June	July	Amount of Change	Percent Change
Net sales	$15,400	$16,800		
Cost of goods sold	$ 9100	$10,010		
Gross profit on sales	$ 6300	$ 6790		
Operating expenses	$ 3610	$ 5140		
Net income	$ 2690	$ 1650		

6. You are the proprietor of Trans-Freight Systems. Your income statements showed the following figures. Complete the table.

	Last Year	This Year	Amount of Change	Percent Change
Net sales	$216,480	$234,170		
Cost of goods sold	$108,240	$117,085		
Gross profit on sales	$108,240	$117,085		
Operating expenses	$ 52,610	$ 54,180		
Net income	$ 55,630	$ 62,905		

Copyright © by Glencoe Division

Name ______________________________ Date ______________

The Plaza Five: A Simulation

You are the manager of "The Plaza Five," a group of five businesses located in the City Squires Shopping Plaza. Complete the table to find the monthly rental charge for each business.

Business	Number of Dimensions	Annual Rate Square Feet	Monthly per Square Foot	Rental Charge
Glamour Boutique	40 ft by 40 ft		$10.50	
Comfort Shoes	50 ft by 40 ft		$11.20	
Modern Barbers	50 ft by 55 ft		$10.75	
Sure-Cut Jewelers	80 ft by 105 ft		$11.40	
Quick Stop Grocery	150 ft by 180 ft		$ 9.85	

You will pay for the maintenance and improvement jobs shown below. Complete the table to find the total charge for each job.

Job	Time Required	Number of Employees	Hourly Rate	Labor Charge	Materials Charge	Total Charge
Painting	40 h	4	$18.75		$1147.50	
Installing Hair Dryers	8 h	4	$14.50		$ 924.80	
Carpeting Shoe Store	$4\frac{1}{2}$ h	6	$10.50		$2745.00	
Installing Security System	12 h	2	$15.00		$1245.00	

Complete the table to find the total monthly cost of telephone service for each business shown below. Round to the nearest cent.

Business	Basic Monthly Charge	Additional Local Calls	Cost per Additional Call	Cost of All Additional Calls	3% Federal Excise Tax	Total Cost for Month
Glamour Boutique	$25.40	15	$0.08			
Sure-Cut Jewelers	$25.40	12	$0.08			
Comfort Shoes	$24.80	32	$0.08			
Modern Barbers	$24.80	14	$0.08			
Quick Stop Grocery	$27.50	9	$0.08			

Copyright © by Glencoe Division

Name ________________________ Date ____________

The Plaza Five: A Simulation
(CONTINUED)

What is the total monthly cost of telephone service for Sure-Cut Jewelers when the federal tax rate is 2%? ____________

What is the total monthly cost of telephone service for Comfort Shoes when the federal excise tax rate is 0%? ____________

The Plaza Five used a total of 56,800 kilowatt-hours of electricity in December. The peak load for the month was 125 kilowatts. Use the chart shown to find the cost of electricity for the month.

What was the cost of electricity for the month? ____________

NORTHWEST POWER	
Demand Charge	
First 50 kW	$6.54/kW
Over 50 kW	$5.91/kW
Energy Charge	
First 250 kW•h	$.1055/kW•h
Next 750 kW•h	.0954/kW•h
Next 2000 kW•h	.0644/kW•h
Next 2000 kW•h	.0578/kW•h
Next 5000 kW•h	.0488/kW•h
Over 10,000 kW•h	.0455/kW•h
Fuel Adjustment	$.017/kW•h

You apportion the total expenses among the five businesses. The amount that each business is charged depends on the space that it occupies. Complete the table to find the amount each business pays.

Business	Number of Square Feet	Total Square Footage	Total Expense	Amount Paid
Sure-Cut Jewelers		41,750 square feet	$12,400	
Comfort Shoes		41,750 square feet	$12,400	
Glamour Boutique		41,750 square feet	$12,400	
Quick Stop Grocery		41,750 square feet	$12,400	
Modern Barbers		41,750 square feet	$12,400	

FINAL THOUGHT

The location a business occupies in a mall or shopping plaza can help to determine the amount of service or merchandise the business sells. Some businesses may pay a higher rental charge than others depending on the attractiveness of their locations. What are some of the factors that make one location more attractive than another?

____

____

Copyright © by Glencoe Division

Name ____________________ Date ____________________

LESSON 21-1

Corporate Income Tax

Taxable income is the portion of your company's gross income that remains after normal business expenses are deducted. The structure of federal corporate income taxes is graduated.

TAXABLE INCOME = ANNUAL GROSS INCOME − DEDUCTIONS

Refer to the table on page 532 (21-1) of your textbook for federal corporate income tax rates.

	1.	2.	3.	4.
Corporation	Miller, Inc.	Hot Wok	J. T. Newton	KMJ
Annual Gross Income	$216,750.00	$ 116,418.00	$67,800.00	$217,400
Deductions	$ 98,415.00	$ 51,420.00	$24,950.00	$125,350
Taxable Income				
Total Tax				

5. Pappas Manufacturing Company had these business expenses for the year:

Wages	$516,450.00	Property taxes	$174,196.00
Rent	$ 48,000.00	Depreciation	$ 38,750.00
Utilities	$ 13,960.00	Other deductions	$ 14,614.90
Interest	$ 6,247.85		

Pappas had a gross income of $1,826,000 for the year. What is the total of Pappas' business expenses? ____________

What is Pappas' taxable income? ____________

What is Pappas' federal corporate income tax for the year? ____________

6. Dr. Alice McFee formed a medical corporation. The corporation had these business expenses for the year:

Wages	$56,500.00	Property taxes	$ 4,749.00
Utilities	$ 7,840.00	Depreciation	$14,916.50
Interest	$ 6,000.00	Other deductions	$ 3,500.00

The corporation's gross income for the year was $154,850. What is the federal corporate income tax for the year? ____________

7. Your business had a gross income of $874,600 for the year. You deducted the following business expenses from the gross income:

Wages	$574,600.00	Interest	$ 9,800.00
Utilities	$ 10,630.00	Depreciation	$20,294.00
Rent	$ 48,000.00	Other deductions	$11,430.00

What was the total for business expenses? ____________

What was the taxable income? ____________

What was the federal corporate income tax for the year? ____________

Copyright © by Glencoe Division

Name ____________________ Date ____________

LESSON 21-2

Issuing Stocks and Bonds

When your company issues stocks or bonds, you must pay certain expenses. The amount that your business actually receives from the sale after paying theses expenses is the net proceeds.

NET PROCEEDS = VALUE OF ISSUE − TOTAL SELLING EXPENSES

---

1. Pilgrim Utility Company issued 20,000 shares of stock at $35 per share. Find the net proceeds after these selling expenses are deducted. ____________

Underwriting Expenses		Other Expenses	
Commissions	$42,000	Printing costs	$14,800
Legal fee	6,000	Legal fees	23,000
Advertising	4,700	Accounting fees	15,600
Miscellaneous	2,800	Miscellaneous	5,000

2. Global Corporation sold 400,000 shares of stock at $27.25 per share. The investment banker's commission was 6.3% of the value of the stock. The other expenses were 0.6% of the value of the stocks. What net proceeds did Global Corporation receive? ____________

3. Riverside Development Co. sold 250,000 shares of stock at $28.75 per share. The underwriting commission was 5.4% of the value of the stocks. The other expenses were 0.9% of the value of the stocks. What net proceeds did Riverside Development receive? ____________

4. The Tokay Fund is composed of 500 investors who invested $3000 each. The Fund will be distributed as shown. What are the dollar amounts for each item?

	Percent	Amount
Gross Proceeds	100%	____________
Expenses:		
Underwriting commission	12.5%	____________
Acquisition fees	10.0%	____________
Capital reserves	7.5%	____________
Net Proceeds		____________

5. You own an engine manufacturing firm. You have plans for a major expansion. To finance the program, you plan to sell 500,000 shares of stock at $40.00 per share.

	Percent	Amount
Gross Proceeds	100%	____________
Expenses:		
Underwriting commission	6.3%	____________
Accounting fees	0.4%	____________
Legal fees	0.3%	____________
Printing fees	0.2%	____________
Miscellaneous expenses	0.1%	____________
Net Proceeds		____________

Copyright © by Glencoe Division

Name ______________________________ Date ______________

LESSON 21-3

Borrowing

Your business may take out a commercial loan to buy raw materials, products, or equipment. The maturity value of your loan is the total amount you repay. Commercial loans usually charge ordinary interest at exact time.

MATURITY VALUE = PRINCIPAL + INTEREST OWED

---

1. Maine National Bank loaned $58,000 to White Mountain Lumber Co. The term of the loan was 180 days. The interest was 10.5%. What was the maturity value of the loan? ______________

2. North Industries wanted to purchase the stock of a company that was going out of business. Federal National Finance agreed to loan North $260,000 for 60 days. Federal National charged 9.75% interest. What is the maturity value of the loan? ______________

3. To take advantage of a special medical supplies sale, College Pharmacy borrowed $17,400 from the City Trust Company. City Trust charged 12.4% interest on the loan. The term of the loan was 135 days. What was the maturity value of the loan? ______________

4. Southwest Bank charges exact interest while Sunshine Trust Company charges ordinary interest. Clark Fox plans to borrow $28,500 for 75 days at 11.8%. What is the maturity value if the loan is from:

a. Southwest Bank ______________ **b.** Sunshine Trust ______________

5. In order to meet its July payroll, Acme Battery Company borrowed $18,500 on July 15 at 9.8% ordinary interest at exact time. The due date of the loan is September 25. What is the maturity value of the loan? ______________

6. The Open Door Company needs to borrow $380,000 for 180 days to help finance the production of a new model door. The business manager arranged financing from 2 sources. Each loan charges ordinary interest at exact time.

Georgia Trust Company	**Peachtree Investment Company**
$134,000 for 180 days Interest rate: 11.5%	$246,000 for 180 days Interest rate: 11.25%

What is the total interest for the 2 loans? ______________

What is the total maturity value? ______________

7. Your company needs $480,000 for 320 days to help finance the development and production of some computer software. You arranged the financing from these 3 sources. Each charges interest as indicated.

Jefferson Trust	**Washington Investments**	**Lincoln Investment Bankers**
$160,000 at 11.4% Ordinary interest at exact time	$150,000 at 11.75% Exact interest at exact time	$170,000 at 12% Ordinary interest at exact time

What is the total interest for the loans? ______________

What is the total maturity value? ______________

Copyright © by Glencoe Division

Name ______________________ Date ____________

LESSONS 21-4, 21-5

Investments: Cost of T-Bill, Net Interest

Your business may invest surplus cash by purchasing United States Treasury Bills. These bills are issued on a discount basis. The face value of the Treasury Bill is the amount of money you will receive on the maturity date of the bill. Your business can also invest in a 30-day to 270-day commercial paper, issued by a company having a high credit rating. Both investments earn ordinary interest at exact time.

COST OF TREASURY BILL = (FACE VALUE OF BILL − INTEREST) + SERVICE FEE

NET INTEREST = TOTAL INTEREST EARNED − SERVICE FEE

1. The City Opera Company had a very successful season. They purchased $20,000 in commercial paper for 30 days from City Investment Corporation. The interest rate was 8.5%. The bank charged a $20 service fee. What net interest will the opera company receive from this transaction? ____________

2. Sable Motor Car Corporation offered commercial paper at 7.95% for 30- to 89-day notes and at 8.1% for 90- to 119-day notes. How much more interest do you earn on a 90-day $1000 note rather than an 89-day $1000 note? ____________

3. The financial manager of Great Lakes Marina made the following investments. The bank charges a service fee of $25 on notes less than $30,000. What is the total net interest for each loan?

Western States Copper Mine	$25,000 at 8.4% for 40 days	____________
Tennessee Lumber Co.	$40,000 at 8.75% for 90 days	____________

4. Your company has just received $400,000 for the sale of a warehouse. As the financial manager, you have to decide the best way to invest the money. You are considering whether to invest in a Treasury Bill or in commercial paper.

U.S. Treasury Bill	City Corporation
Interest rate 4.5%	Interest rate 4.65%
Bill matures in 91 days	Time: 90 days
Bank service fee $40	Bank service fee $40

How much interest will you earn from the Treasury Bill? ____________

How much interest will you earn from the commercial paper? ____________

5. On July 1 your company received $340,000 for the sale of real estate. As financial manager you invested the money in commercial paper. The bank service fee is $40 on all notes less than $40,000. What is the total net interest?

Link Securities	$ 60,000 at 6.5% until July 29	____________
J. S. Best & Co.	$100,000 at 6.75% until Aug. 1	____________
Thomas Morriss & Sons	$150,000 at 7.12% until Sept. 10	____________
Axelrod-Timkins Ltd.	$ 30,000 at 7.08% until Oct. 24	____________
	Total net interest	____________

Copyright © by Glencoe Division

Name ______________________________ Date ______________

LESSON 21-6

Growth Expenses

You may expand your business in many ways. The cost of expansion may include construction fees, consultation fees, legal fees, and so on.

TOTAL COST OF EXPANSION = SUM OF INDIVIDUAL COSTS

1. The Far East Restaurant is expanding its kitchen. They pay $3500 for construction cost and $987 to restock utensils. How much did they pay for this expansion? ______

2. Ashton Architects, Inc. plans to open a new branch office. Ashton purchased property for $32,800. Construction costs for a new building totaled $425,700. In addition, Ashton paid an architect's fee of 13.1% of the cost of construction. Legal fees totaled $5000. New equipment and fixtures cost $12,450. Other expenses came to $4500. What is the total cost of expansion? ______

3. Your company is expanding by adding a new department. You are converting an area of 1800 square feet into offices. The costs of expansion are:

Construction permit	$45
Removal of 3 walls	$1143
New lighting fixtures	$1575
New office equipment	$6425
New carpeting at	$11.40 per square yard

What is the total cost of expansion for your company? ______

4. Your computer software firm is planning to expand its business by building a new building. Expenses include:

Land	$ 60,000
Building construction	1,250,000
Architect's fee 9.5% of construction cost	______
Construction manager 5.5% of construction cost	______
Landscaping	8,150
Legal fees	7,500
Equipment and fixtures	84,325
Additional supplies	2,150
Miscellaneous expenses	5,000
Total Growth Expenses	______

5. Silver Realty Company plans to open a new branch office in another city. Growth expenses include:

Consultant fees	$ 8,400
Real estate agent 5% of first year rent (office space cost is $2,500 per month)	______
Interior decorator	2,175
New furniture/fixtures	9,498
Legal fees	1,200
Travel expenses	3,650
Total Growth Expenses	______

Copyright © by Glencoe Division

Name ______________________ Date ______________

LESSON 22-1

Inflation

Inflation is an economic condition during which there are price increases in the cost of goods and services.

The inflation rate is expressed as a percent increase over a specified time period, usually to the nearest one-tenth of one percent.

$$\text{INFLATION RATE} = \frac{\text{CURRENT PRICE} - \text{ORIGINAL PRICE}}{\text{ORIGINAL PRICE}}$$

The current price can be found by adding the original price to the inflation rate times the original price.

CURRENT PRICE = ORIGINAL PRICE + (ORIGINAL PRICE × INFLATION RATE)

The original price can be found by dividing the current price by one plus the inflation rate.

ORIGINAL PRICE = CURRENT PRICE ÷ (1 + INFLATION RATE)

	Inflation Rate	Current Price	Original Price
1.		$147.79	$ 119.90
2.		$ 27.98	$ 24.99
3.	3.5%		$ 75.49
4.	5.4%		$18,500.00
5.	0.8%	$ 14.29	
6.	12.5%	$978.65	

7. A new auto that currently sells for $25,000, sold for $24,000 one year ago. Find the inflation rate for autos for that one year. ______

8. Find the current price of a home that sold for $100,000 two years ago if the inflation rate for homes over that two-year period is 10%. ______

9. Find the original price of a lawnmower that currently sells for $595 if the inflation rate for lawnmowers for that period is 5%. ______

10. In 1938, one pound of coffee cost $0.17. Currently, one pound of coffee costs $2.99. What is the inflation rate for coffee from 1938 to the present? ______

11. One year ago unleaded regular gasoline was selling for $0.988 a gallon. The inflation rate for gasoline over the past year is 15%. What is the current price for a gallon of unleaded regular gasoline? ______

12. Find the original price of an outdoor grill that currently sells for $179.98 if the inflation rate for grills for that period is 8.6%. ______

Copyright © by Glencoe Division

Name ______________________ Date ______________

LESSON 22-2

The Gross National Product (GNP)

The *Gross National Product (GNP)* is a measure of a nation's economic performance. The GNP is the estimated total value of all goods and services produced by a nation during a year. Only goods and services that add to the national income are included; do-it-yourself activities are not.

A nation's GNP can appear to be growing faster than it really is because of inflation. For this reason, the *real GNP,* or adjusted GNP, is corrected for inflation:

Real GNP = GNP − (Inflation Rate × GNP)

The *Per-capita GNP* can provide an indication of the nation's standard of living. It is the GNP distributed over the population:

Per-Capita GNP = GNP ÷ Population

	GNP	Inflation Rate	Population	Real GNP	Per-Capita GNP
1.	$120.0 billion	5.0%	40.0 million		
2.	$ 96.0 million	13.4%	480,000		
3.	$956,500,000	1.2%	235,000		
4.	$ 24.6 billion	0.4%	2.1 million		
5.	$478.6 million	7.5%	32.4 thousand		
6.	$13.4 trillion	3.6%	1.2 billion		

7. A country has a population of 60,000,000; an inflation rate of 4.0% and a GNP of $300 billion. Find

a. the Real GNP ______

b. the Per-capita GNP ______

8. France has a gross national product of $943 billion. The population of France is 56,184,000, and the inflation rate of France is 3.5%. Find

a. the Real GNP ______

b. the Per-capita GNP ______

9. The GNP of Italy is $825,000,000,000, with a population of 57.657 million. The rate of inflation in Italy is 6.2%. Find

a. the Real GNP ______

b. the Per-capita GNP ______

10. In 1990, the United States had a GNP of $5.2 trillion and a population of 248.7 million. The rate of inflation for that year was 4.6%. Find

a. the Real GNP ______

b. The Per-capita GNP ______

Copyright © by Glencoe Division

Name _______________________________________ Date ______________

LESSON 22-3

Consumer Price Index (CPI)

The *Consumer Price Index (CPI)* is a measure of the average change in prices of a certain number of goods and services. The year 1967 is used as the base year and the CPI for 1967 is set at 100. To find the CPI for any given commodity divide the current cost by the cost in 1967 and multiply by 100 (round to the nearest tenth):

CPI = (CURRENT COST ÷ COST IN 1967) × 100

If you know the CPI for a given commodity and its cost in 1967, you can find the current cost by multiplying the cost in 1967 by the CPI and then dividing by 100:

CURRENT COST = (COST IN 1967 × CPI) ÷ 100

If you know the CPI for a given commodity and its current cost, you can find the cost in 1967 by dividing the current cost by the CPI and multiplying by 100:

COST IN 1967 = (CURRENT COST ÷ CPI) × 100

		Consumer Price Index (CPI)	Current Cost	Cost In 1967
1.	New home		$106,800	$ 42,720
2.	Groceries		$ 48.00	$ 15.25
3.	Dining out	212.8		$ 23.50
4.	New suit	186.3		$190.00
5.	New car	312.7	$18,750	
6.	Gasoline	364.6	$ 1.099	

7. The present cost of a lawnchair is $75.00. The cost in 1967 was $37.50. Find the CPI to the nearest tenth. ______

8. The cost in 1967 of a good pair of gym shoes was $34.75. The CPI for gym shoes is 170. Find the present cost. ______

9. The present cost of a classical cassette is $12.00. The CPI for classical cassettes is 200. Find the cost in 1967. ______

10. Tickets for the new musical on Broadway are selling for $65.00 each. In 1967, the same tickets would have cost $19.75. What is the CPI for recreational services such as theatre tickets? ______

11. In 1967, the Browns paid $50 per month to heat their house. The CPI for heating oil is 504.8. What would it cost the Browns today to heat their house for a month? ______

12. A haircut at the local franchise haircutting salon costs $8.50. The CPI for personal care services such as haircuts is 170. What would a haircut have cost in 1967? ______

Copyright © by Glencoe Division

Name ______________________ Date ______________

LESSON 22-4

Budget

A *budget* enables a business, governmental agency, or an individual to identify what sources are expected to produce revenue (earn money) and what amounts are allocated to various departments or categories for expenses. Revenue and expenses are assigned as a percent of the total amount. The actual amount spent must be compared with the budget amount.

BUDGET ALLOCATION = PERCENT × TOTAL INCOME

DIFFERENCE = ACTUAL AMOUNT − BUDGET ALLOCATION

In problems 1–6, find the budget amount and the difference.

	Total	Percent	Actual	Budget	Difference
1.	$ 60,000	20.0%	$ 11,000		
2.	$ 150,000	30.0%	$ 42,500		
3.	$ 825,000	40.0%	$ 295,000		
4.	$ 4,650,000	17.5%	$ 825,000		
5.	$ 8,548,750	21.8%	$1,645,960		
6.	$37,986,900	5.4%	$1,978,425		

7. Turnkey Limited has a $600,000 budget. For each category determine the amount budgeted and the difference.

Total	Percent	Actual	Budget	Difference
Salaries	78.8%	$465,780		
Supplies	7.5%	$ 46,500		
Equipment	6.5%	$ 39,000		
Maintenance	5.0%	$ 27,860		
Miscellaneous	2.2%	$ 20,860		
Total				

8. Allocate $2,450,000 in revenue as shown.

Sales . . . . 75% __________ Service . . . . 18% __________ Investments . . . . 7% __________

9. Mulholland Manufacturing Co., Inc. is budgeting total revenues next year of $12,575,000. Out of the total, 90% is expected to come from sales, 5% from service contracts, 3% from investments, and 2% from miscellaneous sources. How many dollars has Mulholland budgeted in expected revenues in each category?

____

10. The Mulholland Manufacturing Co., Inc. in problem 9 had actual revenues of $11,450,000 from sales, $598,600 from service contracts, $407,800 from investments, and $232,500 from miscellaneous sources. Did Mulholland reach its revenue goals?

____

____

____

____

Copyright © by Glencoe Division

Notes

Notes

Notes